KINGSLAND
The Mermaid
HOMERTON
Balls Pond
Newington Green Lane
Kingsland Road
Rhodes Farm
White Lead Works
Mutton & Cat
Haggerstone
Sheep Lane
Scotts Tile Kilns
Cambridge Heath
Goldsmiths Alms Ho.
Frame Knitters Alms Ho.
Ironmongers Alms Ho.
Drapers Alms Ho.
Durham Pl.
Edith Pl.
Old Ford Lane
OLD FORD
Drift Way
Hermitage
Twig Folly
Madhouse
Globe Lane
Bancroft's Hospital
Mile End Road
Ducking Pond
North St.
Mile End
to Poplar
London Hosp.
Rhodes Well
St. Dunstans Church
STEPNEY
Salmon Lane
Commercial Road
LIMEHOUSE
Cable Street
New Road
Moor Fields
Bethlem
London Docks
ROTHERHITHE
Bason
Mill Dam
Commercial Docks & Ponds
Cuckold's Point
Limehouse Hole
Deptford Road
Greenland Dock
Bermondsey Spa
Grange Road
Islands
Blue Anchor Road
Bricklayers Arms
Paragon
Hatch
China Hall
Plough Lane
Red Lion
Manor R.
St. Helena
Midway Place
Deptford Lower Road
Victualling Office
Dock Yard
Street Farm
Glue Manufactory
Halfpenny Hatch
Surry Square
The Dover Road
Walworth Common
GRAND SURRY CANAL
Water Works
DEPTFORD
2 Miles
DON.
Jones Ave Maria Lane.

City of beasts

Manchester University Press

CITY OF BEASTS

How animals shaped Georgian London

Thomas Almeroth-Williams

Manchester University Press

Published by Manchester University Press
Altrincham Street, Manchester M1 7JA
www.manchesteruniversitypress.co.uk

British Library Cataloguing-in-Publication Data
A catalogue record for this book is available from the British Library

ISBN 978 1 5261 2635 1 hardback

First published 2019

Typeset by Newgen Publishing UK
Printed in Great Britain
by Bell & Bain Ltd, Glasgow

Contents

Illustrations

Preface

The endless stream of men, and moving things,
From hour to hour the illimitable walk
Still among streets with clouds and sky above,
The wealth, the bustle and the eagerness,
The glittering Chariots with their pamper'd Steeds,
Stalls, Barrows, Porters: midway in the Street
The Scavenger, who begs with hat in hand,
The labouring Hackney Coaches, the rash speed
Of Coaches travelling far, whirl'd on with horn
Loud blowing, and the sturdy Drayman's Team,
Ascending from some Alley of the Thames
And striking right across the crowded Thames
Til the fore Horse veer round with punctual skill:
Here there and everywhere a weary throng

William Wordsworth, *The Prelude*, Book VII (1805), lines 158–171[1]

Early one morning in 2005, 200 years after this impression of London first appeared in print, an unfamiliar sound stirred the residents of Kentish Town. Peering between the curtains, I watched as about thirty horses walked calmly along Prince of Wales Road, together with a few camouflaged riders. For perhaps ten seconds, as their hooves clattered on the tarmac, the windows rattled. They passed out of sight, then earshot and before long, the din of internal combustion engines restored normality to the street. I later discovered that the horses were returning to St John's Wood Barracks, the then headquarters of the King's Troop Royal Horse

Artillery.[2] This proved to be a one-off encounter and apart from occasionally seeing a mounted policeman near the Bank of England, horses never encroached on my urban experience again. The peculiarity of meeting large four-legged animals in a major modern city like London, and the fact that this has the power to shock the senses, unsettle, delight and linger in the memory, is significant. It is becoming increasingly difficult for Londoners, and urban dwellers in many other cities, particularly in the West, to imagine a time when horses, cattle, sheep, pigs and other livestock were ubiquitous and essential presences in their streets. But this is precisely the London that William Wordsworth encountered when he visited in the late eighteenth century.

The hustle and bustle of the Georgian metropolis, which amazed Wordsworth as much as it had Samuel Johnson a generation earlier, provides rich pickings for historians, novelists and film-makers.[3] In this period, London became the most populous, prosperous and powerful city in the world – between 1715 and 1815, the city's population increased from 630,000 to 1.4 million – and as countless studies have emphasised, this new kind of metropolis fulfilled myriad roles: global trade hub, fashionable playground, transport nerve centre, social melting pot and poverty trap among them.[4] Historians of Georgian London are fortunate because groundbreaking projects have made the city one of the most digitised in human history. Since the launch of Old Bailey Proceedings Online in 2003, researchers have gained rapid access to nearly 200,000 trials from London's central criminal court, providing insights into millions of metropolitan lives; and in 2010, the London Lives project digitised and made text-searchable 240,000 manuscript and printed pages, containing more than 3.35 million name instances, from eight London archives.[5] These resources are invaluable and foster ambitious new approaches to historical research, but we may fail to do them justice if we continue to follow overly narrow lines of established enquiry. Of particular concern, in the context of this book, is the risk that by encouraging searches for human names and lives, digitisation will unintentionally contribute to the traditional portrayal of cities as being the product of human activity alone.

Preface

A City Full of People, the title of Peter Earle's survey of London between 1650 and 1750, and a phrase borrowed from Daniel Defoe, neatly conveys the extent to which the story of cities has been told as the story of people.[6] Very few historians have acknowledged the city's animals and even fewer have integrated them into key debates in social, urban and economic history. I started researching this book after discovering that animals were almost entirely absent from an 836-page volume of the *Cambridge Urban History of Britain* covering the period 1540–1840, as well as three of the most popular biographies of London.[7] Over the past ten years, I have picked at a seemingly inexhaustible seam of evidence of animal life and influence in the city. This book draws on a wide range of sources, but as a pig farmer's son and an advocate of writing history from below, I unapologetically prioritise evidence of tangible, dung-bespattered interactions between real people and real animals. In doing so, I deliberately steer away from the theoretical and fictional sources that continue to dominate English animal studies. As a result, this book offers new insights into the lived experiences of Georgian Londoners, as well as the workings and character of a city about which we still have so much to learn. I aim to reveal the extraordinary contribution that certain four-legged animals made to the social, economic and cultural life of the world's first modern metropolis, as well as the serious challenges which they posed. Following in the hoof- and paw-prints of these beasts, this book reappraises London's role in the industrial and consumer revolutions, as well as the city's social relations and culture. In doing so, it calls for animals to be set free from the pigeon-hole of 'animal studies', to be accorded agency and to be integrated into social and urban history.

Acknowledgements

As an undergraduate in Durham many years ago, it took a while to find a period or topic which really inspired me. Finally, in Adrian Green's lecture series about everyday life in the eighteenth century, something clicked. In the four years after graduating, the need to earn a crust made it difficult to keep doing history but Simon Fowler and Penny Lodge kindly allowed me to publish features about Georgian England in the National Archives' *Ancestors* magazine, one of which was entitled 'City of Beasts'. From then on, I couldn't shake the idea for this book, even when I was supposed to be thinking about press releases for frozen ready meals, art exhibitions and access to higher education. I am grateful to office-based friends, past and present, for tolerating and encouraging me.

In 2008, I gained the support of an extraordinary supervisor, Mark Jenner, who began to nurture the genesis of this book at Master's and then PhD level at the University of York. Like the most skilful of livestock drovers, Mark has coaxed and goaded my unruly thoughts through a labyrinth of challenges. For everything he has taught me, as well as his consistent enthusiasm and kindness, I will always be grateful.

York's Centre for Eighteenth Century Studies offered a supportive environment while I was studying, even when I was many miles away, and kindly made me a research associate in the final stages of writing this book. I am particularly grateful to Hannah Greig for sage and inspiring advice, scholarly and parental; but also to Alison O'Byrne, Helen

Cowie, Sarah Goldsmith and Emilie Murphy. Outside of York, many brilliant scholars have given me valuable guidance and encouragement, not least Robert Shoemaker, Chris Pearson, Peter Guillery, Jon Stobart, Ian Bristow, Derek Morris, David Turner, Louise Falcini and Charlie Turpie. Above all, Roey Sweet and Tim Hitchcock have been tremendous long-term allies. Any remaining faults are, of course, entirely of my own making. In the end, Durham's greatest gift was not a degree but a room-mate nonpareil and I am relieved to be able (finally) to thank Simon Willis for spurring me on for so long. The same goes for Owen Mason, another faithful university friend, who made sure that I stayed afloat in 2017.

This book is deeply indebted to the teams behind the Old Bailey Proceedings Online and London Lives – in particular Tim Hitchcock, Robert Shoemaker, Clive Emsley, Sharon Howard and Jamie McLaughlin – but also to countless librarians and archivists in London, Kew, Windsor and Sheffield. I am especially grateful to the superb staff of the London Metropolitan Archives, where I spent my happiest and most productive hours.

I am grateful to everyone who has listened to me give papers and asked questions, especially attendees of the Institute of Historical Research's 'Long Eighteenth Century' seminar and Leicester's Centre for Urban History. My thanks also go to the Curriers' Company and the *London Journal* for the support I received in the London History Essay Prize of 2017. Writing this book has consumed money as well as my youth, and would have been impossible without an AHRC PhD studentship. Cambridge University Press have kindly permitted me to republish research in Chapters 1 and 2 that first appeared in 'The brewery horse and the importance of equine power in Hanoverian London', *Urban History*, 40 (2013), 416–41. I am also grateful to Taylor & Francis for allowing me to republish, in Chapter 7, parts of 'The watchdogs of Georgian London: non-human agency, crime prevention and control of urban space', *London Journal*, 43 (2018), 267–88.

Converting a dream into printed reality can be a bewildering process so I am fortunate that my editor and the wider team at Manchester

University Press have put their trust in me and treated this project with such enthusiasm. I am also grateful to the anonymous readers for their helpful comments at both proposal and draft manuscript stages.

More than anyone else, I would like to thank my family, starting with my amazing parents, Lesley and Derek Williams, who imparted their love of history and nature, among countless other secrets to a fulfilling life. Without their support and that of my grandparents, Horace 'Bill' and Ruby Almeroth, I doubt that I would have even considered writing this book.

When shown a room-sized copy of an old map labelled with animal hotspots, most dates would have run a mile, but Hélène Almeroth-Williams (née Frélon) moved in and has lived with the ups and downs of this obsession ever since. I never could have dreamt that one person could give me so much love, strength, inspiration and joy as you do, Hélène. I dedicate this book to you and to our wonderful children, Joseph and Julia.

Abbreviations

AgHR	*Agricultural History Review*
BL	British Library, London
BMDPD	British Museum, London, Department of Prints and Drawings
EcHR	*Economic History Review*
LL	London Lives: www.londonlives.org
LMA	London Metropolitan Archives
OBP	Old Bailey Proceedings Online: www.oldbaileyonline.org
TNA	The National Archives, Kew

1 John Rocque, *Rocque's Plan of London and its Environs* (engraving, *c.* 1741).

Introduction

Georgian London spearheaded Britain's Enlightenment ambitions to tame and turn a profit from Mother Nature. It was from the Port of London that ships traversed the globe to acquire plants, timber, animal skins and live exotic beasts; and it was from the metropolis that Britain schemed to transplant livestock and crops from one continent to another to serve the imperial economy. In the 1780s, George III extracted merino sheep from Spain and nurtured this precious flock in the gardens of Kew and Windsor to produce breeding stock for the improvement of English wool. In 1804, the progeny of these animals were dispatched to the British colony of New South Wales and the seeds of a major industry were sown. Meanwhile, botanists, merchants and government ministers began to consider the possibility of cultivating Chinese tea plants in the mountains of Assam and Bhutan.[1] Along every trade route, Britain tightened its grip on nature but nowhere was this more striking than in the metropolis itself. In 1830, William Cobbett described London as the 'all-devouring WEN', a monstrous force stripping the countryside of people, livestock and grain.[2] Cobbett associated London's consumption of animal lives with rural poverty, but it was also an awe-inspiring demonstration of Britain's growing power and prosperity.

This book reveals a city of beasts which has been hiding in plain sight. A basic but often overlooked feature of William Hogarth's *Second Stage of Cruelty*, 1751 (see Figure 2), one of the most iconic images of Georgian London, is that animals are as prevalent as people. Hogarth, who was born in the shadow of Smithfield Market in 1697, depicts sheep, horses, a bullock and a jack-ass swarming into a Holborn cul-de-sac. As numerous scholars have observed, the artist uses the ensuing melee to expose the laziness, greed and cruelty of his fellow Londoners, but the city's relationship with animals was far more complex than this might suggest, and never stood still.[3] By the time William Wordsworth visited in 1788, the intensity of horse traffic was far greater than anything Hogarth had known and by the early 1800s, it was estimated that 31,000 horses were at work in and around the metropolis. At the same time, around 30,000 sheep and cattle were driven through the streets to Smithfield Market every week. No other settlement in Europe or North America, in any earlier period, had accommodated so many large four-legged animals, or felt their influence so profoundly. And no other city in the world can provide more compelling evidence for John Berger's assertion that before the twentieth century, animals were 'with man at the centre of his world'.[4]

This book is about London but it also seeks to challenge a lingering tendency to view all cities, past, present and future, as being somehow divorced from the influence of animals, an assumption that threatens to exaggerate their artificial characteristics and downplay their complex relationship with the natural world. Since the early 1990s, scholars of North America, in particular, have led the charge for urban environmental history and opened up new avenues of research to consider the role played by animals in social and urban history.[5] As we continue down a path of accelerating urbanisation and eco-crisis, there has never been a more important time to consider what cities have been, what they are today and what they could be in the future. Historians and social scientists have long disagreed about what cities represent and where their boundaries lie.[6] Since the 1980s, there has been a growing emphasis on 'unbounding' cities in various ways to conceptualise them as 'spatially open and connected'.[7] Bruce Braun observes that 'urbanization occurs in and through a vast network of relationships, and within complex flows of energy and matter, as well as capital, commodities, people and ideas, that link urban natures with distant sites and distant ecologies'; while Samuel Hays calls for a consideration of 'the direct interface between the city and

2 William Hogarth, *Second Stage of Cruelty* (etching and engraving, 1751).

the countryside'.[8] Thinking about urban animals sheds new light on this debate: as we will see, Georgian London's interactions with and impact on animals extended far beyond the geographical area upon which this book focuses, that is the more than 40 km^2 surveyed by John Rocque in the late 1730s and early 1740s. This comprises the Cities of London and Westminster, the Borough of Southwark, and suburban zones including Chelsea, Bermondsey, Deptford, Stepney, Shoreditch and Clerkenwell (see Figure 1). Before we set off, it is important to keep in mind that in the early 1800s, less than two-thirds of this area was built-up and that market gardens, orchards, fields and meadows continued to occupy at least 4,000 acres of what was recognised to be part of the metropolis. We will occasionally venture further afield, but only to examine interactions between people and animals when they were travelling to or spending a few hours away from London. Nevertheless, this book seeks to blur the traditional boundaries of 'town' versus 'country', and 'urban' versus 'rural' because, as Roy Porter asserted:

> Man has made the country no less than he has made the town, and from this it follows that the historical relations between town and country are contingent, expressions in part of changing images of the urban and the pastoral… The comparative history of urbanism is an enticing field, or rather piazza, ripe for further study.[9]

One of the many challenges facing urban environmental histories and urban nature studies is deciding whether it is appropriate to conceive of a city as a unified or consistent whole.[10] Historians have often commented on the diverse functions that London performed in the Georgian period, including its role as the heart of government, justice and the royal court; as well as being a hub of banking, trade, consumption, sociability, art and publishing.[11] But there has been a tendency to carve London into four contrasting parts: the West End, the City, the East End and Southwark. This is now changing, in no small part because the ongoing digitisation of Georgian London's archival records is helping historians to focus in more closely as well as trace complex patterns throughout the city.[12] This book contributes to this process, because the activity and influence of particular animals in one street could change dramatically in the next. By tracking their hoof- and paw-prints, I hope to show that Georgian London was a complex weave of variegated urban topographies, land uses and social types.

The study of animals in historical contexts has evolved dramatically over the last thirty years. The combined effect of the rise of environmental history and the 'cultural turn' since the 1970s has freed animals from their traditional home in agricultural-economic geographies and allowed them to roam across the humanities and social sciences. Looking at British history more specifically, however, the publication of Keith Thomas' classic *Man and the Natural World* in 1983 was a pivotal moment.[13] Thomas' ambitious assessment of man's relationship to animals and plants in England from 1500 to 1800 firmly established non-human animals as a subject worthy of historical enquiry and remains a scholarly *tour de force*. At the same time, some of Thomas' arguments have provoked criticism. Most importantly, in the context of this book, Thomas claimed that by 1800 English urban societies had become alienated from animals, observing that the rise of new sentimental attitudes was 'closely linked to the growth of towns and the emergence of an industrial order in which animals became increasingly marginal in the processes of production'. Thomas acknowledged that working animals were 'extensively used during the first century and a half of industrialization' and that horses 'did not disappear from the streets until the 1920s' but 'long before that', he claimed,

> most people were working in industries powered by non-animal means. The shift to other sources of industrial power was accelerated by the introduction of steam and the greater employment of water power at the end of the eighteenth century; and the urban isolation from animals in which the new feelings were generated dates from even earlier.

London represents the most advanced model for Thomas' hypothesis: here, above all other cities, he would expect to find 'well-to-do townsmen, remote from the agricultural process and inclined to think of animals as pets rather than as working livestock'.[14] My research tests these assumptions and challenges conventional urban historiographies by exploring Georgian London as a human–animal hybrid, a city of beasts as well as a city 'full of people'.[15] My argument is not just that animals occupied the city in force, it is that they underpinned its physical, social, economic and cultural development in diverse and fundamental ways.

I am not the first to question the idea that animals were peripheral in eighteenth- or nineteenth-century London, but in reassessing the relationship between animals and English society, previous studies have overwhelmingly focused on issues of animal cruelty and the rise of humanitarianism. Their central aim has been to show, in contrast to Thomas' view, that 'it was not philosophical distance from sites of cruelty, but painful proximity to them which prompted Londoners' protests'.[16] While this approach is valuable, the tendency to consider human–animal histories as narratives of abuse also threatens to oversimplify complex relationships and the context in which they were formed. When considering the treatment of animals in this period, we have to remember that this was a city in which infants regularly died before they could walk; petty thieves were hanged or transported to penal colonies; servants and apprentices were violently abused; and children performed dangerous manual labour. The victim model has also led historians to neglect the multifaceted roles that animals played in Georgian society and to downplay their ability to make things happen. While, for instance, several scholars have discussed the ill-treatment of horses, there has been scant analysis of the economic significance of equine haulage, its impact on the construction and use of metropolitan space, or the challenges of commanding equine behaviour. Part of the problem has been that animal studies relating to England from 1500 to 1900 have tended to rely on theoretical sources, particularly philosophical and religious works; natural histories and Romantic literature.[17] Many of those who produced this commentary viewed urban life from afar or had little or no personal experience of working with animals. Thus, while they reveal a great deal about animal symbolism, anthropomorphism, Romanticism and other developments in intellectual history, they do not tell us very much about tangible interactions between real people and real animals.[18] Some historians have begun to challenge and depart from this approach – Ingrid Tague, for instance, has emphasised 'the importance of lived experiences' to 'remind us that pets were not merely metaphors used to think about the world but living, breathing beings that had a direct impact on the lives of the humans with whom they interacted'[19] – but there is much more work to be done.

As in other areas of historical enquiry, the cultural turn has guided animal studies into privileging, as John Tosh put it, 'representation over experience' and eliding 'social history and its quest for the historical lived experience'.[20] One of the most striking consequences of this has been, as Tim Hitchcock complained in 2004, that academic history has largely abandoned 'the experience of the poor' to focus on 'the words of the middling sort' and the 'glittering lives of the better off'.[21] By contrast, this book is rooted in the rich seam of evidence generated by those who had first-hand experience of the urban beast, including the plebeian men, women and children who lived and worked with animals, as well as the magistrates, beadles, constables and watchmen who sought to regulate their behaviour on the streets. Instead of searching for the emergence of modern London 'between the ears of the middling sort', this book traces it through the dung-bespattered interactions that Londoners had with animals in the city. The key characters in this narrative are, therefore, not clergymen, writers or politicians, but London's brewers, brickmakers, tanners, grocers, cow-keepers, coachmen, horse dealers, drovers, carters, grooms and warehousemen. Scholarly neglect of these largely plebeian Londoners goes some way to explaining why the city's animals have also been so overlooked. Horses and livestock in Georgian England spent most of their lives with low-born workers but generally only attract attention when they were being ridden by, admired or painted for the elite. By foregrounding the city's animals, therefore, this book hopes to give further momentum to the movement for social history from below.

While historians are now giving elite women attention, the lives of other female Londoners remain in the shadows. With the exceptions of cow- and ass-keeping, the occupations most closely associated with animals were almost exclusively held by men. Inevitably, therefore, this book reveals more about the working lives of men than it does those of women. Milkmaids are briefly discussed, as are female pig-keepers, but more importantly, this book shows that all female Londoners experienced and helped to create the city of beasts. While elite women rode or travelled in horse-drawn carriages, for instance, female pedestrians had to weave between horses to avoid being soiled, kicked or run over. At the same time, female demand for meat, milk, leather and diverse manufactured goods helped to fuel the city's insatiable consumption of animal flesh and power.

Looking for evidence of human–animal interactions has given me the opportunity to draw lesser-known Londoners, including large numbers of low-paid workers, into the light for the first time. I have unearthed evidence from an array of sources including commercial, legal and parliamentary records; newspaper reports and adverts; diaries and personal correspondence; maps and architectural plans; paintings, prints and sketches. To begin to do justice to the lives of the people and animals recorded in this material, I hope to shift the focus of historical enquiry away from debates centred on intellectual history, the rise of kindness, humanitarianism and animal welfare legislation; towards the integration of animals into wider debates about urban life. In doing so, I want to reassess what Georgian London was, what the city was like to live in, how it functioned and what role it played in some of the major developments of the period.

Previous studies of this city have given the impression that the presence of animals was incongruous with the key manifestations of the capital's success in this period: thriving commerce, grand architecture and the fashionable lifestyles of polite society. In doing so, some historians have presented animals as generic case studies of nuisance. Emily Cockayne has, for instance, considered how people living in England from 1600 to 1770 'were made to feel uncomfortable' by the 'noise, appearance, behaviour, proximity and odours' of other beings. In Cockayne's survey of English towns, including London, pigs are reduced to 'notorious mobile street nuisances', dogs are condemned for barking and biting, and horses associated with producing copious amounts of stinking dung as well as being involved in accidents.[22] These impressions echo the horror and disgust expressed by mid-nineteenth-century social and sanitarian reformers as they sought to cleanse London, Manchester and other cities of animal life in the name of human progress and urban improvement.[23] Yet, such a one-sided approach threatens to underestimate early modern urban governance, caricature the challenges posed by animal behaviour, ignore the positive contribution these actors made, and neglect the complexities of human–animal relationships.

City of Beasts challenges the dominant view of London's social history as being the product of human agency alone, but what do I mean by agency? Environmental historians have argued for decades that nature has agency but have not always made it clear what this entails. Animal historians have theorised non-human agency more thoroughly but some, together with a number of environmental historians, have been tempted to describe it as 'resistance'. This is problematic because 'resistance' is loaded with political meaning and deeply entwined with human psychology, but also because conflating animal recalcitrance, aggression and other behaviours with resistance makes an anthropomorphic assumption that animals are conscious of their oppression and can envisage an alternative future, as well as how to bring this about.[24] Furthermore, as Chris Pearson has argued, focusing on resistance immediately erects a barrier between human and non-human agents which conceals their 'close relationship' as well as how 'their ability to act is contingent on these historically situated relations'.[25] These are precisely the barriers which this book aims to break down and, like Pearson, I have found far more effective conceptualisations of agency emerging from the cross-disciplinary movement crystallising around Bruno Latour's work on actor–network theory. This argues that the social is performed by non-human things as much as by humans and that '*any thing*' that makes a difference to other actors is an agent.[26] Some scholars have argued that agency requires reason, intentionality and self-consciousness.[27] But there are compelling arguments against such a rigid position. Research into animal psychology, for instance, supports the idea that the behaviour of some animals can be accorded a degree of intentionality and it is possible to identify examples of this in Georgian London. I would also contend, however, that there are different and, in the context of this study, more influential forms of agency, including those which do not rely on intentionality. I will demonstrate that, despite obvious inequalities in power relations, horses, livestock and dogs were capable of influencing human behaviour in significant ways. As we will see, this included constraining and obstructing human activity but this study is primarily concerned with the ways in which non-human animals empowered, encouraged and made things possible for Georgian Londoners.[28] This is an approach that foregrounds the entwined lives of people and four-legged animals and considers what their quotidian interactions contributed to the evolution of Georgian London.

Animals were so ubiquitous that tracking them down opens up remarkable new perspectives on unfamiliar or misunderstood social types, spaces, activities, relationships and forces which enable us to challenge assumptions about London's economic, social and cultural development. For the first time, non-human life takes centre stage in the major themes of eighteenth- and early nineteenth-century English urban history: commerce, trade and industry; the consumer revolution; urban expansion and improvement; social relations, crime and disorder.

Chapters 1 and 2 challenge two common misconceptions: first, that London played a marginal role in the Industrial Revolution and, second, that steam substituted animal muscle power. Despite recent calls to look beyond a narrow band of technological innovations and to acknowledge the existence of other British industrial revolutions in which human industriousness played a key role, historians continue to focus on steam power and sideline horses. This book offers an alternative to innovation-centric accounts of technological progress and undermines the notion that the supposed 'failure' to substitute new for old is to be explained by conservatism, lack of ambition or ignorance.[29] These opening chapters demonstrate that London's dependence on equine muscle power and the co-operation of men and horses increased dramatically in the Georgian period. Chapter 1 reveals that the mill horse helped to transform industrial production in the metropolis long before the introduction of Boulton and Watt's groundbreaking Sun and Planet type steam engine in the 1780s, and remained an effective power source in some trades well into the nineteenth century. Chapter 2 considers the contribution which the city's draught horses made to the metropolitan economy as manufacturing and international trade boomed, generating unprecedented demand for haulage by road. It reveals that the city's draught horses were valued for their intelligence as well as for their power, and explores the relationship between human and non-human co-workers as the challenges which they faced, side-by-side, intensified.

If Georgian London has been peripheral in orthodox studies of the Industrial Revolution, the city's bearing on the agricultural revolution has been made to appear equally obscure. We have been led to believe that

Georgian Londoners were agriculturally unproductive, an impression that sits all too comfortably with twenty-first-century Western expectations of how a civilised city must function. Urban societies, particularly those in developed nations, are becoming increasingly alienated from the source of their food but Georgian London presents a very different form of urbanity at the start of the modern age, in which livestock were a familiar feature of the urban environment, and their relationship to consumers much more intimate. Chapter 3 examines the role played by urban cow- and pig-keeping in feeding the metropolitan population and shows that these activities adapted to urbanisation and industrialisation rather than becoming their victim in this period. Nevertheless, urban husbandry was far from able to satisfy the capital's voracious appetite for animal flesh. In a period predating refrigeration and railways, London relied on most of its meat being delivered alive and on the hoof from the English, Welsh and Scottish countryside by drovers. Chapter 4 reveals that the Smithfield livestock trade was a major sector of the metropolitan economy but also that its operations impacted on the lives of all Georgian Londoners. Smithfield's location meant that thousands of sheep and cattle had to be driven back and forth across the city, a system which not only brought disorder to the streets but also maximised the population's exposure to and interaction with these animals.

London's influential role in the consumer revolution has been studied from many different perspectives but the city's demand for animals has received remarkably little attention. In addition to its consumption of cows, sheep and pigs, the capital exerted a powerful draw on Britain's equine stock, and dominated the trade in riding and private coach horses. By 1800, thousands of these animals were sold in the city every year in a thriving economic sector that promoted innovative commercial practices. Chapter 5 examines the evolution and cultural significance of this trade but also proposes that these horses were voracious consumers in their own right and, therefore, demonstrated significant agency in the consumer revolution. Having established that riding and carriage horses were expensive and troublesome to maintain, Chapter 6 considers why so many Londoners opted to invest in them. The capital's recreational life has been the subject of numerous studies but these generally give the impression that assemblies, balls and concerts dominated the season and that Londoners only derived pleasure from sociability. Meanwhile, historians tend to associate the quintessentially Georgian pursuits of horse riding, racing and hunting with the countryside. I challenge these assumptions, revealing that London was the mainspring of British equestrian culture and that its residents dedicated huge amounts of time, money and energy to their animals. By studying tangible interactions between horse and rider, it becomes clear that the city's equestrian culture both facilitated sociability and offered an alluring alternative to human company.

Closely linked to rising consumption was the problem of property crime and a new way of exploring this is through London's relationship with dogs. Theft was a major concern in the Georgian period and, as numerous studies have shown, this led to major developments in crime prevention and punishment. But when thinking about dogs in Georgian England, historians have tended to focus on nuisance curs or cossetted lapdogs, giving the impression that the species was a hindrance to or a distraction from prosperity and police. Yet, as Chapter 7 reveals, dogs fulfilled a significant role in the metropolitan economy by guarding valuable property against thieves. In doing so, it challenges the impression that access to private space, a key battleground in Georgian power relations, was controlled solely by humans; and argues that the presence and behaviour of watchdogs shaped the urban experience of thousands of people.

City of Beasts invites readers to explore the Georgian metropolis and its population in a new light. This book does not set out to provide an encyclopaedia of animal life in the city. Instead, it focuses on the animals which a wealth of evidence suggests had the greatest impact. As a result, it largely omits birds, insects, fish, wild mammals (including rodents), stray and feral animals, as well as imported birds and beasts. Each deserves attention elsewhere and recent studies by Christopher Plumb and Ingrid Tague have already emphasised the cultural significance of exotics.[30] Yet, these creatures were, on the whole, caged objects of display, a condition which greatly restricted their interactions with people. This book concentrates on the animals that Georgian Londoners themselves viewed as being most 'useful' in their city; that is, horses, cattle, sheep, pigs and dogs. These were the animals that powered, fed and guarded the metropolis; but they also walked its streets and interacted, in one way or another, with the entire population.

1

Mill horse

London is generally viewed as a bit player in Britain's Industrial Revolution. The dominant view remains that the capital was slow to embrace steam power and remained technologically conservative; while banking, insurance and overseas trade, rather than manufacturing, defined the capital's position in the national economy.[1] In the 1990s, however, some historians began to argue that London was an industrial powerhouse and that its leading manufacturers were pioneers in the use of steam-powered machinery.[2] There is clear evidence to support this: as early as 1733, the city employed seven steam engines for pumping water and by 1780 this had increased to twenty-seven. More importantly, as soon as Boulton and Watt invented a steam engine capable of powering diverse manufacturing work, London became a leading investor. The company's first double-acting engine was sold to the East Smithfield brewer Henry Goodwyn in 1784 and over the next twenty years, the capital was second only to Lancashire in the acquisition of the Sun and Planet type engine.[3] And yet, this entire debate is blinkered, reinforcing a fixation on coal and steam. Britain's discovery of abundant coal seams and invention of technologies able to convert this fuel into mechanical power played a crucial role in the Industrial Revolution.[4] But there were other fundamental drivers of change – political, economic, geographical and social – which historians have only just begun to explore, and others that continue to be overlooked. It is all too tempting to focus on the contribution of luminaries like Richard Arkwright and Josiah

Wedgwood but they would have been hamstrung without the skill and hard work of millions of men, women and children. Led by Jan de Vries, a growing number of historians now argue that the Industrial Revolution relied on an 'industrious revolution' and there is convincing evidence to suggest that labour input increased substantially in England from the mid-eighteenth century. Craig Muldrew, for instance, has shown that agricultural workers began to consume a much more calorific diet in this period and so had more energy to burn in the fields. And Hans-Joachim Voth has demonstrated that in London, traditional ways of life were restructured to make more time for work.[5] These are important revisions but why stop at human labour?

This chapter widens the concept of an 'industrious revolution' to include interactions between men, animals and machinery. In doing so, it argues for a less innovation-centric approach to the Industrial Revolution. As David Edgerton has argued, a technology's continued use over the centuries does not make it any less effective or any less able to complement radical new technologies and processes. In modern Britain, horse work is associated with a bygone rustic age, occasionally remembered in folk museums and country fairs, but largely sidelined from economic and urban history.[6] By contrast, this book situates horse power at the very heart of London's Industrial Revolution. In doing so, it calls for a broader ecological perspective on urban and industrial history. As we will see, Georgian Britain plundered much more than coal from Mother Nature and staged a momentous episode in the history of human interaction with horses. The Industrial Revolution integrated horse work into economic systems in radically new ways, prompting major advances in selective breeding – to achieve particular specifications in size, speed, strength and stamina – as well as in the training and care of horses. Horse–human co-operation proliferated in virtually every part of the British Isles, in its villages, ports and manufacturing towns; on its farms and coal fields, along its canals and newly turnpiked roads, and in its forests and quarries. And yet, it was in the metropolis that horses made their single greatest contribution to the economy. Nowhere else were working horses concentrated in such vast numbers or employed in so many influential industrial activities. To bring these previously overlooked aspects of London life into view, historians need to sally forth from the Royal Exchange, the bank, the counting house, the coffee shop and other familiar haunts into the relative *terra incognita* of London's manufactories, warehouses, streets, stables and wharves. In these locations, new dimensions of the metropolitan economy become clear, throwing into question long-accepted assumptions about the city, its role in the Industrial Revolution and the nature of technological progress.

Steam propelled British manufacturing into the modern age, ending centuries of inertia and facilitating mass production for the first time, or so we have been led to believe. These assumptions permeate orthodox narratives of the Industrial Revolution but ride roughshod over the chronology of technological change and overlook a complex tapestry of progress. Consider the vertical watermill, an ancient technology which drove spectacular economic progress in Georgian England. The country boasted well over 10,000 by the early 1700s and by the end of the century, waterwheels generated most of the power required by one of Britain's most advanced manufacturing sectors, the textile industry. Between 1770 and 1815, cotton production soared by an astonishing 2,200 per cent but this was predominantly achieved by traditional waterwheels rather than steam-powered technology.[7] At the same time, systems that thrived in one part of the country floundered elsewhere. In London, waterwheels played a minimal role in industry because, as well as being subject to tidal activity, the Thames discharged far too much water over too shallow a gradient to make use of the most efficient system. The overshot model simultaneously harnessed the weight and momentum of flowing water as it plunged into buckets from above, but the only significant metropolitan business to benefit from this system was the London Bridge Water Works (1581–1822), which managed to suspend waterwheels from the arches of the bridge and adjust their position in harmony with the tide. This proved highly effective – by the early nineteenth century, five wheels were supplying about four million gallons of water per day to 10,000 customers – but it was virtually impossible to replicate elsewhere in the city. Some manufacturers made do with the least efficient type of waterwheel, the undershot, which rotated as water flowed against its submerged base. In Southwark, these tide wheels dated back to the medieval period and at least one, at St Saviour's Dock, remained in operation for much of the eighteenth century. At high tide, sluice gates admitted Thames water into a system of mill-ponds; then at low tide, the gates were opened and the stored water released, creating sufficient flow to turn a wheel. This technology served small-scale production well enough but competition for land in the eighteenth century rendered such mills increasingly uneconomical.[8]

The city fared little better with wind power, being low-lying and sheltered by hills. The New River Company constructed a six-sailed windmill in Islington in the early 1700s to pump water to its new reservoir but this soon had to be replaced with a horse-wheel because unlike the wind, horses supplied power on demand. A few suburban windmills, including three on the Surrey side of the Thames, survived into the second half of the eighteenth century but urban sprawl gradually swept them away.[9] As one Londoner belatedly explained in 1856: 'Our busy city cannot now spare room for windmills; and if there were such room, the wind could barely get at them.'[10] Unable to exploit steam before 1784, manufacturers had to find an alternative source of power and the resource which London amassed in astonishing abundance in this period was muscle power, both horse and human. Perhaps more than in any other part of Britain, London's Industrial Revolution relied on the collaborative labour of men and horses; and the linchpin of this partnership was an ancient technology: the mill. In its basic form, horse mills, also referred to in this period as an 'engine', or 'gin' for short,[11] comprised a vertical rotating drive shaft, horizontal beams – to which the mill horses were harnessed – and a circular horse walk. But as we will see, the addition of gears, millstones, pumps, borers, rollers and other mechanisms adapted these engines for diverse and innovative manufacturing work.

Mill horses lack obvious appeal as heroes of Georgian England. Often blind and nearing the end of their working lives, they were certainly not fine physical specimens. Almost any medium-sized horse in command of its legs could do the job and many mill horses were rejects from the uncompromising hackney and stagecoach trades. These animals propped up the equine hierarchy and their lives were sufficiently gruelling to elicit pity from at least some Londoners by the late eighteenth century. In 1785, 'The High Mettled Racer', a song performed on the London stage, lamented the decline of a used-up racehorse. Still well known in the 1830s, the song inspired an engraving after Robert Cruikshank in 1831 (see Figure 3), which shows the animal 'Blind, old, lean, and feeble' tugging 'round a mill' under the watchful eye of its unfeeling owner.[12] As moved as some theatregoers may have been, however, such sympathetic attitudes were far from representative of metropolitan society. We can learn much more about perceptions of horses from responses to technological progress. Eighteenth-century London was a wellspring and enthusiastic consumer of prints, models, lectures, books and newspaper articles which actively celebrated horse work and presented mill horses without sentiment but as vital components of sophisticated mechanical processes. In 1748, for instance, the Holborn engraver William Henry Toms published an intricate engraving of a horse-powered machine recently used to drive piles

3 George Wilmot Bonner from designs by Robert Cruikshank, untitled wood engraving published in Charles Dibdin, *The High Mettled Racer* (1831).

for the construction of Westminster Bridge, which opened in November 1750. A meticulous key explained that 'By the Horses going round, the great Rope is wound about ye Drum, & the Ram is drawn up till the Tongs come between the inclin'd Planes; where they are opened & the Ram is discharg'd'. The machine incorporated a 'Trundle head with a Fly', a device specially designed by the watchmaker James Vauloué 'to prevent the Horses falling' into the Thames 'when the Ram is discharged'. In *A Description of Westminster Bridge*, published the following year, its architect, Charles Labelye, boasted that 'by the Force of three Horses going at a common Pace' this engine had generated five strokes every two minutes.[13] We should not allow this kind of rational scientific commentary to disguise or diminish the harsh reality that horses were routinely worked to death in Georgian London, as well as beaten, malnourished and otherwise neglected. Nevertheless, to present horses as mere victims of human exploitation belittles their achievements and denies them agency.

It is difficult to get to know mill horses in the archives because descriptions of individual animals and their daily lives are so rare. An intriguing exception appears in a City of London coroner's inquest deposition from 1795. That year, the Batemans, a successful family of Clerkenwell silversmiths, hired paviours to improve the company's stables. To get to this part of the manufactory, the men found that they had to scurry between two horses working a flatting mill, described further below. One paviour made it through but his co-worker was crushed by the rotating pole of the horse-wheel and then trampled by the animals. A groom soon arrived, stopped the horses 'going round with the Mill' and the horse-keeper brought a light, only to find that the man was already dead. This case does not tell us anything about the physical or behavioural traits of the individual horses involved but it does offer a sense of the repetitive nature of their labour, as well as the dingy and confined spaces in which they toiled. It also emphasises that these animals worked alongside people, that they were periodically left unsupervised, and that they placed significant demands on human labour.[14]

Mill horses helped the Batemans to become one of Britain's leading manufacturers of silverware but the trade was a relatively minor consumer of equine power, remaining heavily reliant on handwork. These animals made a far greater contribution in several other metropolitan industries. Evidence from sale-of-stock newspaper adverts, insurance policy documents and the Old Bailey Proceedings make it possible to locate, in time and place, more than fifty horse-powered gins employed across twelve different trades for the period 1740–1815.[15] This only represents a fraction of the total in operation at any one time; as we will see, by the 1780s, the brewing industry alone employed at least fifty horse gins, but this sample offers valuable information about the application of horse power in particular industrial contexts. As shown in Figure 4, London's equine engines were concentrated in Old Street, Clerkenwell, Shoreditch, East Smithfield, Southwark and Bermondsey, the city's principal manufacturing districts. The vast majority were used to process raw materials, or more specifically, for grinding pigments for colourmen and dye-makers; tobacco for snuff-makers; wood and minerals for druggists and perfumers; flint for glass-makers and potters; and most significantly, grain for brewers, distillers and corn-chandlers. Brick-making, another major industry which pockmarked London's northern suburbs, also used horse power to mix and knead brick-earth with water and soot, a process known as pugging. There was remarkably little technological change in the building industry in the Georgian period and until the mid-eighteenth century, pugging was entirely performed by human hands and feet, one of many examples of gruelling labour on brickfields.[16] The horse-driven pug mill was a rare example of innovation and by 1800, most, if not all, substantial operations had made the transition. The trade relied on men, women, children and horses working together on a tightly controlled production line. Brick earth was first dug out with shovels and ferried by wheelbarrow to the pugging mills. Once churned into a malleable clay, this material was extracted from the drum and heaped in front of brick-moulding gangs, often whole families, which were expected to turn out hundreds of bricks every day. Finally, 'taking-off boys' loaded the wet bricks onto barrows and wheeled them to the drying hacks.[17] Horse work on the brickfields was strenuous and monotonous but no more so than that experienced by

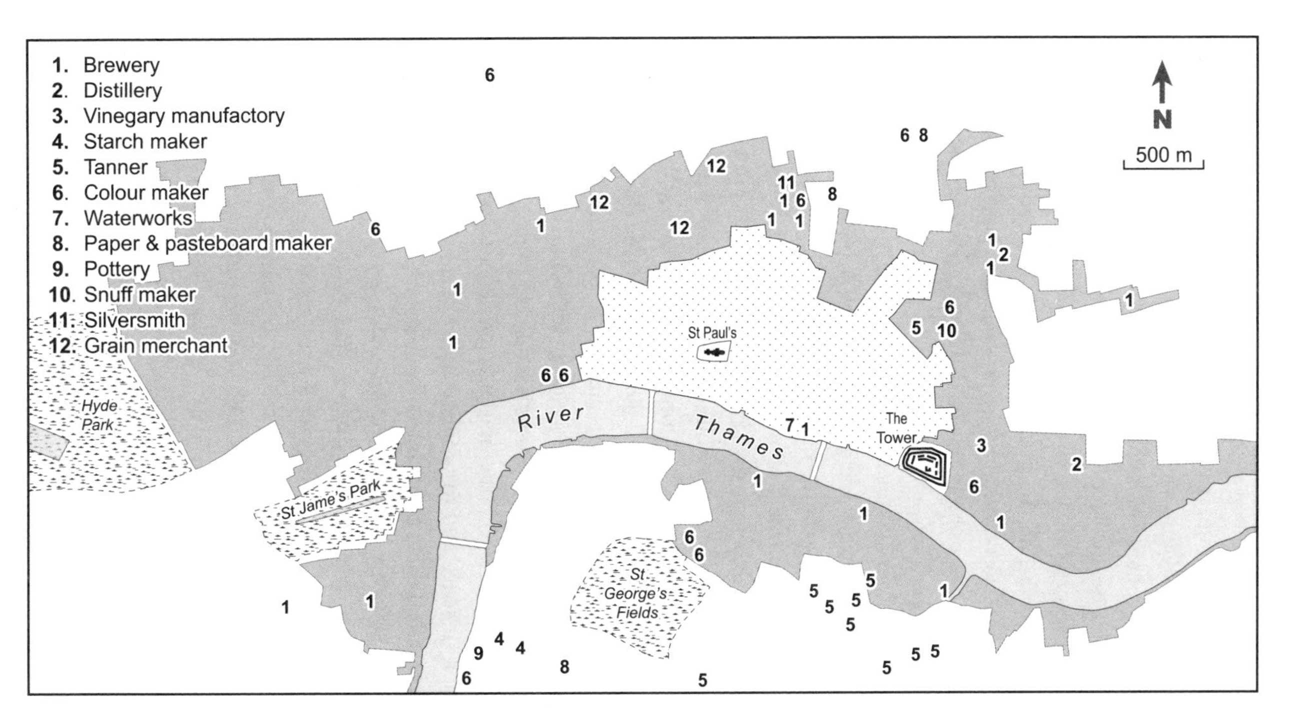

4 Distribution of horse mills in London, 1740–1815.

their human co-workers. Co-operation of this kind not only permeated London's economic life, it was intrinsic to the city's industrial success. To find evidence of this, we need only follow our noses to Southwark, London's pungent leather-making district.

Aided by a ready supply of raw materials, tanners flourished in eighteenth-century London. The metropolitan livestock trade, discussed in Chapter 4, delivered the finest hides at a competitive price and with minimal cartage costs; London's river, wells and pipe network supplied water; and the city's shipping and haulage infrastructure ensured the availability of bark, the source of tannin. At the same time, tanners could depend on local demand from the city's leather finishing trades. By the 1820s, curriers, leather-cutters, dressers, dyers, saddlers, harness- and collar-makers, glovers, hatters, cordwainers and boot-makers accounted for around 1,000 firms and about 7 per cent of London's manufacturing businesses.[18] Nevertheless, the city's tanners faced a growing threat in this period. As with brick-making, tanning required relatively little plant and machinery, instead relying on human strength and manual skills. But paying men to perform unskilled tasks became increasingly uneconomical. As tanners sought to reduce wage costs and expand production, equine technology came to play a vital role. Tanners needed large quantities of bark to extract enough tannin to transform raw hides into leather. Ripped from oak trees in the English countryside, bark was sent to London in strips and to produce a tannin solution or 'ooze', tanners first had to grind these down into dust. Use of horse-powered bark mills in England probably began in the seventeenth century when the rotary grinding mill or edge runner, which had existed in parts of Europe since ancient times, was being applied to a growing number of manufacturing processes. When harnessed to a rotating drive shaft, a horse giving traction caused a vertical millstone to roll around a pit, crushing and grinding the bark laying in its path (see Figure 5).[19] Occasionally, men would return to rake the bark and whips were almost certainly used to galvanise sluggish animals, but for significant periods of time these animals could be left to work unsupervised. Thus, horse mills promoted efficiency and task specialisation, both hallmarks of the Industrial

5 Anon., *A Tanners Work-shop*, engraving published in *The Universal Magazine of Knowledge and Pleasure* (June 1751).

Revolution: horses took over the most monotonous, heavy work of the tannery, which allowed the human workforce to focus on more complex tasks including scraping hides, preparing vats and administration.[20] Insurance records show that by the 1790s, most if not all of Southwark's tanners were insuring mill houses and stables on their premises. These were modest buildings but the technology within became increasingly sophisticated in this period.[21] Five patents for improved bark mills were issued between 1797 and 1805, each approving the use of horses. The first and most innovative was that patented by James Weldon of Lichfield and consisted of a cylindrical wooden case fitted with a cast iron cone which rotated at speed. Armed with long triangular teeth at the top and shorter ones below, it first chopped the bark coarsely and then reduced it to powder. Four years later, James Whitby introduced a rival system which relied on cutting wheels and when 'moved by a horse', could grind 150 kg of bark per hour. Finally, Thomas Chapman, a Bermondsey skinner, patented a new mechanism which hybridised these systems to offer both a revolving barrel and metal rag-wheels.[22] This new generation of bark mills operated much faster than the traditional edge-runners and these developments undermine the assumption that tanners shunned technological change before the 1830s, when they adopted new steam-powered machinery. In London, technological innovation preceded the adoption of steam engines and it did so supported by tried and trusted equine labour.[23]

Horses, horse mills and stables became important markers of the capitalisation of the tanning trade. In the late eighteenth century, firms began to insure their premises for more than £1,000 due to unprecedented investment in warehouses, workshops, stables, machinery, livestock and stock. Among those leading the charge was George Choumert, a tanner and 'Spanish leather dresser' based in Russell Street, Bermondsey. In 1777, Choumert insured his buildings and their contents for £900; by 1798 this had leapt to more than £11,000, and by 1821 his assets – which included a horse mill and stable containing £100 worth of horses and associated utensils – were valued at £27,000. Some small-scale production continued: in 1817, a tannery on the Camberwell Road was insured for just £500, horse mill included; but by then, a handful of elite firms was producing most of London's leather and for the most part they did so using horses rather than steam.[24] Despite the fact that patents for new bark mills generally approved the use of steam engines, the latter's availability after 1784 had remarkably little impact on the trade. In 1805, John Farey found that only two London tanners were using Boulton and Watt engines, and their power equated to just four horses each. This was partly because very few firms, not even Choumert's, reached a scale of production in the early nineteenth century which made it profitable to buy, fuel and maintain a steam engine. This mirrors a national picture in which uptake of steam engines was limited to mining, textile manufacture, brewing and a few other industries which operated on an unusually large scale. It is important to recognise, however, that horses were also retained because they served certain industrial tasks so effectively. For the majority of tanners the horse remained a reliable and adaptable power source and this reputation influenced industrial decision-making across the country. Even in the West Midlands, where steam engines made early inroads, many manufacturers continued to use horses, as well as wind, water and manpower because they were cheaper and more effective in a particular context.[25]

Far from being a mark of backwardness, mill horses were closely associated with technological and commercial progress in the Georgian period and astute businessmen sought to capitalise on this in their marketing. Few were more successful in this regard than the city's colourmen, or house paint manufacturers. Demand for brightly coloured paints exploded across England but particularly in the wealthy metropolis, which dominated both production and consumption. Several new pigments became available at a price low enough to be applied to walls rather than artists' canvases, Prussian blue (*c.* 1710), Patent Yellow (*c.* 1770), Scheele's Green (1775) and Copper Chloride Green (*c.* 1795) among them. By 1750, it was common for the plaster walls of elite and middling houses to be painted but Palladian interiors, characteristic of the period 1715–55, tended to feature monotone walls within a fairly limited palette. In the 1750s, colour spread to ceilings but this was humdrum compared to the riot which lay ahead. Between 1765 and 1792, Robert and James Adam unleashed an astonishingly enlivened palette and ornate neoclassical designs on the interiors of wealthy West End homes. One of their gaudiest projects was a drawing room for Northumberland House, completed in the 1770s. After applying a bright verdigris ground to the vast ceiling, the brothers picked out elaborate details in pink, red, Prussian blue, black and gold leaf. This was an extravagant showstopper but countless other West End ceilings, walls, fireplaces and window-shutters were decorated with vibrant combinations of Indian yellow, Cerulean blue, olive green and violet, to name just a few.[26] This revolution in interior design had a profound impact on London's paint trade but construction cycles were just as influential. Between 1775 and 1785, hundreds of terrace houses sprang up in the West End, particularly on the Portland, Portman and Bedford estates, in what would be the greatest housing boom for a century. This gave designers a giant blank canvas and generated unprecedented demand for building materials, as well as paint, wallpaper, furniture and fabric. By the 1820s, the capital supported more than 100 retailers of paint.[27]

Purchased by an expanding bourgeoisie as well as the nobility, paint was a hot commodity in London's consumer revolution, but none of this would have been possible without major advances in production, central to which was the application of horse power.[28] A striking demonstration of this appears in a trade card issued around 1744 by Joseph Emerton, a leading paint manufacturer of the day. In one half of the engraved image, Emerton presents a fashionably dressed lady sitting for her portrait but then he beckons prospective customers behind his shop to unveil a horse mill. It may seem incongruous for an advert for an expensive decorative product to feature horse work. Yet, while the forces of taste and commerce often clashed, Georgian consumers would have read this image favourably. The skills involved in industrial processes were celebrated in everything from art and literature to scientific societies and industrial tourism; and within this culture, horse work was emblematic of cutting-edge technologies and processes, as well as innovative or premium-quality goods.[29] A later trade card, issued by the colour-makers Emerton and Manby in the 1790s (see Figure 6), makes even more of this connection, elevating an already intricate depiction of a mill and its power source with an elegant frame and royal crest. One of the company's horses can be seen turning a large central spur gear which drives at least two smaller gears. We are shown four pairs of grinding stones but there were probably eight in total, arranged in a rectangle. This was an impressive mechanical operation, capable of grinding large batches of raw materials simultaneously, more than justifying its owners' evident pride. Colourmen were not the only manufacturers to publicise their horse work in this way. Around 1760, the Cheapside grocer George Farr included a horse-driven 'Spanish Snuff Mill' in an elegant trade card advertising his teas, coffee, snuffs, rum and brandy. Bearing in mind that Georgian trade cards boldly projected any qualities belonging to products or processes that might attract patronage, we can be sure that these manufacturers viewed

6 William Darling and James Thompson, draft trade card of Emerton and Manby, colourmen (etching and engraving, marked 1792).

their horses as marketing assets. This was not only because mill horses conveyed a firm's efficiency and technological sophistication, it was also due to the fact that horses were so highly esteemed in elite culture that even a lowly mill horse lent an air of respectability to what remained a noisome and noxious industry.[30]

Mill horses served the most energy-intensive procedure in the manufacture of paint: the crushing and grinding of minerals, plants, shells and bone to extract pigments. In some operations, horse power was also used to blend these powdered pigments with oil to produce ready-to-use paint. In the seventeenth and early eighteenth centuries, grinding was usually done by hand, either with a pestle or a manually driven mill. This continued in the relatively small-scale production of expensive artists' pigments but a much more powerful mechanical process was needed to produce house paint. In London, the prime mover in this shift to mass-production was the horse mill. The earliest of these machines appeared in the late 1720s at the manufactory of Alexander Emerton on the Strand. While Emerton neglected to mention his horse mill in newspaper adverts, his wife Elizabeth and brother Joseph did so repeatedly as part of a bitter commercial feud which followed his death. In 1742, Joseph notified the public that his sister-in-law was an 'ignorant Pretender' to his late brother's business and advised them to prefer his nearby manufactory on Norfolk Street. Joseph claimed to have improved his late brother's horse mill so that 'no Person in England can exceed, if equal, him in the Perfection or Cheapness of his Colours'; but a few days later, his rival proclaimed that

> as her Colours are ground in HORSE-MILLS, of which there are not the like in England, they are prepared in much greater Perfection, and Sold considerable Cheaper than by any of the Trade that have not such Conveniences, but grind their Colours in Hand-Mills, or upon a Stone, with great Expence and Labour.[31]

In this competitive trade, horse power became a powerful weapon. Knowing that customers wanted the smoothest possible finish, both companies boasted that their mills ground more 'perfectly', or finely, than manual tools could achieve. At the same time, they promised 'Cheapness' by passing on the substantial saving made by substituting horses for men and by selling mass-produced pigments which customers could mix at home. In 1728, Alexander Emerton boasted that 'Five pounds worth of Colours will paint as much Work as a House Painter will do for Twenty Pounds' and insisted that with the aid of his 'Printed Direction', gentlemen and builders could employ their own servants or labourers, rather than professional house painters.[32] This early example of DIY retailing hit house painters hard. In 1747, the *London Tradesman* observed that their business was at 'a very low Ebb' because 'some Colour-shops... have set up Horse-Mills to grind the colours' so that 'a House may be painted by a common Labourer at one Third the Expence it would have cost before'.[33]

The horse mill severed the link between the manufacture of paint and the business of applying it; some house-painters went bust while others were forced to specialise in more niche services such as gilding, japanning and marbling.[34] Meanwhile, London's colour-makers thrived. By the 1820s, there were at least forty-two trading in the capital, mostly in Southwark, Whitechapel and Old Street.[35] The Emerton rivals kept more central premises than most but they were not out of place; the Strand's polite shops fronted numerous backstreet manufactories in which men and horses laboured side-by-side. Warren's blacking factory, for instance, where a twelve-year-old Charles Dickens worked in 1824, was only a short walk away at Hungerford Stairs. Warren's was one of more than fifty blacking and whiting manufactories in London, the latter using horse mills to grind chalk for whitewashing, the former to break down charred bones to colour shoe polish.[36] Proximity to the West End helped the Emertons to draw wealthy customers to their shops and while their workshops were almost certainly hidden from view, the sounds of clopping hooves, intersecting gears and grating stones must have carried into the shop. This would have been a source of pride and some customers may have been invited to explore the mill house to satisfy curiosity and guarantee the authenticity of their expensive purchases. Sale of stock adverts show that the size and power of colour mills varied between firms but increased over time. In the 1770s, a horse-wheel near St George's Fields powered two pairs of grinding stones; another in Chiswell Street operated four stones; and in Hoxton, a 4.5 m diameter wheel was advertised as being 'large'. Twenty years later, a three-storey manufactory in Whitechapel commanded a 6.4 m diameter horse-wheel which put the owner in the same league as brewers, London's leading industrialists, immediately before they acquired steam engines. By 1805, however, only three colour-makers had ordered Boulton and Watt steam engines and these had a modest average horse-power of nine. For most colour-makers in the early nineteenth century, the power of two horses was sufficient to keep up with growing demand.[37]

Although most of London's mills were used to grind raw materials, horses did power other important manufacturing processes in the city. In the 1740s, equine muscle drove fulling mills for wool dyers and mangles for silk dyers in Southwark and the East End. By the 1780s, horses were also working rag mills and paper presses in Lambeth and Old Street for manufacturers of pasteboard, a ubiquitous packaging material.[38] It was around this time that the Bateman silver workshop in Bunhill Row installed its flatting mill, which remained in use until 1802. This machine mass-produced sheet metal from which the Batemans fashioned, with great economy, everything from wine labels and sugar tongs to tea caddies and snuff boxes. Aided by horses, the firm cornered much of the market in middle class domestic silver in the late eighteenth century.[39] In a heavier industrial context, horse power was equally instrumental in the manufacture of weapons at the Royal Arsenal in Woolwich. Introduced in the early 1770s, state-of-the-art Dutch borers carved the finest gun and cannon barrels and did so powered by horses into the mid-1840s. This work bolstered London's prosperity in the late eighteenth century as the city became increasingly reliant on Britain's military might to defend and expand its global trading interests. But mill horses made an even more vital contribution in the service of metropolitan waterworks.[40]

The early modern period witnessed dramatic progress in London's water supply systems, spearheaded by the construction of the New River between 1609 and 1613. Dug by hand over a period of thirty months, this open canal carried water over a distance of 64 km from the River Lea in Hertfordshire to the New River Head reservoir at Islington. From here, water was directed through a wooden main leading down to St John Street, Smithfield and Newgate, where it branched off east and west to supply a rapidly expanding network of pipes. By 1630, the New River served 1,500 homes in the City of London and by the 1650s, five more mains descended from the reservoir. Impressive as these developments were, they laid the foundations for far more in the eighteenth century, spurred on by growth in the city's population and industrial activity. This involved the construction of several new reservoirs, the introduction of steam-powered pumps and a major expansion of the pipe network beyond the City. The work performed by horses changed dramatically in this period but remained indispensable throughout.[41] Horse-powered water pumps were introduced to London in the 1590s, when the Broken Wharf Water Company began to supply a few houses to the north of London Bridge. This continued into the early 1700s, by which time several other companies were using horses to raise water. Accounts for the Clink Waterworks in Southwark suggest that the firm employed at least five horses and payments to a farrier show that these animals required regular treatment for harness wounds, evidence of the debilitating demands thrust on their bodies. The New River Company replaced its impotent windmill, discussed above, with a horse-wheel in the 1720s and Southwark's Bank-End Waterworks used horse power to raise Thames water until the 1750s, when steam began to take over. The New River Company made this transition in 1767, by which point it was already serving more than 30,000 houses.[42] This might have put mill horses out to pasture but instead their workload ballooned in another department.

Since the Roman period, London's pipe network had been built using earthenware and lead but as demand for pipes surged in the seventeenth century, a faster and cheaper solution had to be found. Even before construction of the New River began, the New River Company was experimenting with wooden pipes and they remained in use well into the nineteenth century. Elm and alder were preferred for their durability underground and when exposed to water, as well as for their relatively low density. Brought to London by ship, each tree trunk was hollowed out using an auger and once laid, the narrower end of one pipe was simply driven into the broader opening of the next. Sometimes an iron hoop was used to brace the point of connection to reduce splitting and while leaks remained a problem, the system proved highly effective for over two centuries.[43] By the 1770s, the New River Company was manufacturing and installing around 3 km of pipes annually, and within thirty years, the network comprised several hundred kilometres of pipe reaching from Marylebone in the West to Whitechapel in the East, and from Islington in the North to the Strand in the South.[44] Cast iron pipes did not appear until the 1810s and at first only spread gradually – as late as 1842, it was observed that 'some wooden pipes are still used for conveying water in London'.[45] We can be sure, therefore, that when the New River Company more than doubled its provision to 66,000 houses between 1750 and 1828, it did so almost entirely with wooden pipes. It would be hard to overstate London's debt to these developments. Water was London's blood and wooden pipes acted as its veins, helping to slake the thirst of a population which grew from about 630,000 in 1715 to around 900,000 in 1800, while also supplying a crucial raw material to major industries such as brewing, distilling, dyeing, tanning and soap boiling.[46]

The manufacture of pipes required skill but also a great deal of energy. It was possible to hollow out tree trunks with hand-powered augers but only at the expense of immense physical effort and uneconomical wages. Mass production required a much more powerful and cost-effective engine. Quite when horses superseded men in this trade is obscured by the outbreak, on Christmas Eve 1769, of an inferno at the New River Company's headquarters which destroyed the firm's minute books covering the critical period 1619–1769. In the late 1730s, William Maitland observed that the firm operated a boring machine on its Bridewell site but neglected to mention whether it was horse-powered. In 1756, a revised edition of his *History of London* confirmed that 'Horse Engines' were employed alongside twenty human borers, which suggests that this was a transitional phase, but over the next few years, horses steadily took over the business of supplying power and freed men to focus on more skilled tasks. The firm's human machine operators would begin by hoisting and chaining a fresh trunk onto a trolley. With the mill horses set in motion, the men then wheeled the trunk into the path of the rotating auger and, with careful direction, carved an even cylindrical hollow straight through the trunk.[47]

Minutes taken in the aftermath of the New River Company fire reveal just how reliant the firm had become on horse power by 1770. Having decided to move operations to a new site in Dorset Yard, the directors immediately ordered their surveyor to reinstall the surviving horse-wheels and borers. Within three months, production was back on track and the company began to erect two more 'Horse boring engine works'. There was no time to lose; London was already in the grip of an unprecedented building boom which supercharged demand for pipes. The company raced to sign up new customers and in November 1770 alone committed to install 740 yards of pipe along Gray's Inn Road to supply sixty-six houses. At the same time, the firm sought to capitalise on London's industrial expansion. Orders flooded in from the thriving manufacturing districts of Clerkenwell, Moorgate, East Smithfield and Whitechapel, where the company's rate collectors classified distillers, soap boilers and dyers as 'large Consumers of Water'. Connecting some industrial clients generated weeks of work for men and horses. In 1770, the New River Company hollowed out and laid 120 yards of pipes to supply Mr Bradley's distillery in Drury Lane, for which he 'paid a Fine and Twenty Pounds per annum'; and a few years later, pipes were laid and cocks installed for Gordon's distillery and Meux's brewery, in Clerkenwell. New clients were not, however, the only source of demand for pipes. Leaks generated constant demand for replacements because the life expectancy of a wooden pipe could fall to as little as four years in some soils. Meanwhile, the introduction of steam engines to pump water enabled firms to supply more distant parts of the metropolis but only if they employed additional mill horses to accelerate pipe production. In 1785, the New River Company ordered a new Boulton and Watt engine to raise water to its principal reservoir, the High Pond in Islington. How many horses the firm employed at this time is unclear but minutes refer to the construction and extension of stables next to the pipe workshops to accommodate more livestock.[48]

Evidence of regular repairs being made to the company's horse-wheels gives a sense of the immense forces which these animals exerted but they had their weaknesses and these posed a serious threat to business. In 1806, the board of directors heard that five of its horses were suffering from 'the disorder now prevalent' in the city, most likely equine influenza or strangles. The company's surveyor promised that the farrier and other workers were paying 'every attention' to the sick animals. With the borers standing idle, he suggested calling in a millwright to make repairs but the directors responded with alarm, knowing that without horse power their stockpile of pipes would soon run out. Horrified by the spectre of lost trade, the board ordered him to immediately hire replacement horses and kicked his maintenance plans into the long grass. This exchange goes to show that well into the 'age of steam', industrial decision-making continued to pivot around horses and their interactions with men and machinery.[49]

We have already seen that tanners and colour-makers retained mill horses long after Boulton and Watt began selling their double-acting steam engine, but this was part of a much bigger picture. According to John Farey, in 1799, the capital employed thirty-six steam engines to power manufacturing machinery, rising to eighty-seven in 1805. By then, steam featured in twenty-four trades but of these, three-quarters employed fewer than five engines in total. Half of the capital's steam engines were concentrated in just four exceptionally energy-intensive trades: brewing, distilling, dyeing and metalworking. London was far from backward in harnessing steam power but its take-up was fragmented; even in the sectors which did convert, only a small proportion of firms had done so by 1805, at which point horse mills almost certainly outnumbered steam engines. Horse power did much more than survive, however, it underpinned major technological and industrial progress. This is particularly striking in the case of brewing.[50]

Mill horse

The largest operation in London's powerful food and drink manufacturing sector, brewing underwent spectacular change in the eighteenth century. Buoyed by soaring metropolitan demand, brewers made unprecedented advances in mechanisation, mass production brewing methods, marketing and distribution. Between 1720 and 1799, the number of firms fluctuated between 140 and 180, declining from the 1750s as the biggest operations began to dominate the market. In 1748, twelve elite firms accounted for 42 per cent of metropolitan production, rising to 85 per cent in 1830. Competition was fierce. When the Calvert brothers, William and Felix, each brewed 50,000 barrels for the first time in 1748, a scramble for supremacy ensued. Samuel Whitbread was among the first to achieve 100,000 barrels in 1776 and had doubled this by the end of the century, only to be overtaken by Barclay's, which surpassed 300,000 barrels in 1815. In size and capital, London's leading breweries were then Britain's biggest industrial operations and would retain this position well into the mid-nineteenth century.[51] Historians have tended to associate this meteoric rise with steam but the trade's first double-acting engine was only installed in 1784, prior to which the mechanical components of all large-scale breweries were powered by men and horses. In 1790, only seven brewers used the new technology. As late as 1807, five leading brewers were still reliant on horse engines. And it was not until 1820, when Elliot's of Pimlico converted, that metropolitan brewing truly committed, meaning that coal's triumph over the mill horse took almost forty years to complete. More importantly, this transition had only begun decades after mill horses had already revolutionised the trade.[52]

Traditionally, brewing's most energy intensive processes – grinding malt, pumping water and drawing liquid wort from the mash ton into the copper – had been powered separately. Some larger operations used horses to process their malt but pumps were almost always worked by strong-armed men. This all changed in the first half of the eighteenth century. By the 1740s, many breweries had installed mill wheels which, when linked to a gearing system, enabled them to mill and pump simultaneously. But this technology would have been redundant without an effective prime mover and for half a century, this remained the circular plodding of horses.[53] Ground plans of London breweries often plot the location of horse mills but I have been unable to find a depiction of one in operation. In Paris, however, Denis Diderot did publish a fine engraving of a brewery *manège* (horse walk) and *moulins* (mills) in the *Encyclopédie*, together with a detailed explanation of its workings (see Figure 7). The brewery's location is unspecified and the precise arrangement of the mill may have differed from that found in London but the accompanying description does make clear that the horse-wheel powered both millstones and water pumps.[54] Mill horses supplied energy at a much higher rate and for longer periods than men could ever achieve. Thus, by increasing efficiency and cutting the cost of human labour, these animals facilitated a surge in production and profits, upon which all further advances in brewing were built. So what changed?

To a large extent, the brewer's mill horse was a victim of its own success. In the second half of the eighteenth century, their employment increased dramatically, so that by 1780 a large concern needed at least twenty, with teams of four working together in shifts. These animals served the mechanical needs of the brewery more than adequately but fatigue, need for rest breaks and vulnerability to illness and premature death meant that they were costly. To raise production levels much higher, brewers had to find a more cost-effective system but steam engines were of no use before 1782, when Boulton and Watt adapted their original pumping capability to turn machinery.[55] Until that point, brewers had no alternative to the horse-wheel. As long as they needed horses to grind malt, a steam engine for pumping alone was neither practical nor cost-effective. Once Boulton and Watt had resolved this, several brewers swiftly placed orders. The engineers discovered that some sites were more challenging than others but usually managed to couple the new engine with an existing horse-wheel without major remodelling of buildings.[56] The transition could bring immediate rewards and in 1786, one of Whitbread's assistants, Joseph Delafield, wrote enthusiastically to his brother of the savings

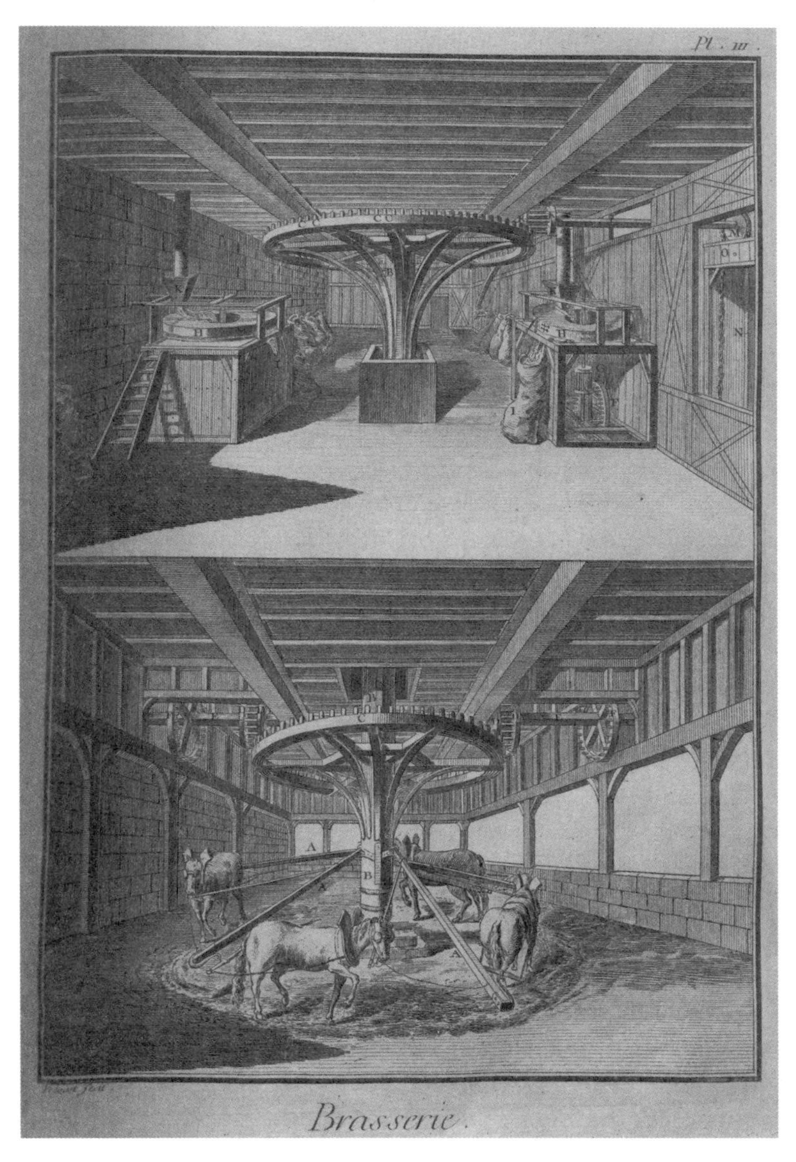

7 Anon., *Brasserie: Planche III: Manège & moulins*, in Denis Diderot and Jean Le Rond D'Alembert (eds), *Encyclopédie ou Dictionnaire Raisonné des Arts et des Métiers*, vol. 19 (Paris, 1763).

they had made within a few months of installation. The mill wheel, he recalled, had

> required 6 horses to turn it, but we ordered our [steam] engine the power of ten [horses], and the work it does we think is equal to fourteen horses, for we grind with all four mills about 40 quarters an hour... We put aside by it full 24 horses, which to keep up and feed did not cost less per annum than £40 a head. The expense of erection was about £1,000. It consumes only a bushel of coals an hour.[57]

Yet, knowing now that steam transformed the trade, we should not assume that brewers found the decision easy to make. Some hard-nosed industrialists let slip an anxious awareness of the risks involved. Barclay Perkins in Southwark was the first to enquire about the Boulton and Watt engine but hesitated before placing an order and was overtaken by Goodwyn, Whitbread and Felix Calvert who were more easily persuaded of its promised advantages. Yet, even Goodwyn, the first to install, retained some of his mill horses at St Kathryn's Dock because he feared the new technology might let him down. In a letter to Boulton and Watt in July 1784, he enthused 'I have parted from one half of my Mill Horses already' but then an air of caution crept in. 'In hopes that you my Engineer, will render them all needless', Goodwyn wrote: 'I am deliberating on the sale of the remainder but shall probably keep two or three until we are perfect masters of the conduct of our new works.'[58] Goodwyn's trepidation was not unusual; in 1790, the Black Eagle in Spitalfields was London's fifth most productive brewery but even this titan of Georgian industry delayed placing an order until December 1807. After seventeen years of vacillation, the firm finally allowed Boulton and Watt to install in the first half of 1808 but even then Sampson Hanbury, who had run the brewery since 1789, proved reluctant to dispense with horse power altogether. A few months before placing his order, Hanbury confidently scaled down from ten to five mill horses, and by the time the engine arrived, only three remained, but then he wavered. Presumably faced with a mechanical glitch, he purchased more mill horses, returning to a total of seven in 1810. The problem took more than a year to resolve but in 1811 the company finally committed fully to steam.[59] These were the experiences of

London's mega-breweries but firms rarely considered installing a steam engine until they had attained 20,000 barrels a year, something which only a quarter of firms managed. In 1805, only 15 per cent of metropolitan breweries owned steam engines, meaning that about 100 operations continued to rely on horse power.[60] Moreover, far from banishing horses from the trade, the surge in production brought about by steam meant that brewers had to expand their haulage operations, and it is to the contribution made by London's draught horses that we now turn.

2

Draught horse

The movement of people and goods by road played a crucial role in the Industrial Revolution.Transport hubs sprang up across England, Scotland and Wales in the eighteenth century, accompanied by major improvements to the road network and unprecedented developments in the organisation of equine labour.[1] Nowhere, however, were horses more impressively mobilised than they were in London. In 1815, it was estimated that 31,000 were employed in Middlesex and a high proportion of these would have drawn vehicles in the capital.[2] This manifested itself as an endless stream of large, powerful animals traversing the city's main arteries. One November morning in 1827, Francis Place sat at his window overlooking Charing Cross and swiftly counted 102 horses heading towards Parliament Street. These included six animals drawing a waggon loaded with turnips, twenty-eight leading seven stagecoaches, twelve hauling three coal waggons, seven straining to move a heavy block of stone, and a pair of dray horses conveying beer.[3] The rapid expansion of this kind of traffic in the Georgian period necessitated the construction, widening and improvement of numerous streets and roads. One of the most ambitious schemes was a 5 km bypass between Paddington and Islington, completed in 1756 (see Chapter 4), but equally significant was the destruction of the old city gateways a few years later. Ludgate, Cripplegate, Aldgate, Aldersgate, Bishopsgate and Moorgate were removed between 1760 and 1762, followed by Newgate in 1777. And the primary purpose of these street improvements was to make it more convenient for horse-drawn carts and waggons to pass through the city's commercial and financial hub.[4]

Since the early twentieth century, draught horses have come to symbolise a bygone rustic age, but throughout the Georgian period, they were essential and influential features of London's bustling urban terrain. A remarkable letter published in *Lloyd's Evening Post* in 1758 sets the scene. Addressed to Parliament and ascribed to the Citizens of London, it proclaims:

> To any man who knows anything of the manner in which Trade is now carried on, it is self-evident that, to a great number of Tradesmen, a Horse is as absolutely necessary as a shop or warehouse. There are also 800 Hackney Coachmen… 420 free Carmen… [and] an innumerable number of Higlers, &c who cannot carry on their business without Horses. Now to all these honest industrious poor People, the raising the Price of Hay is equally oppressive with the raising the Price of Bread.[5]

As this chapter will show, draught horses were at the very heart of the metropolitan economy and underpinned the livelihoods of thousands of people.

More and more Londoners began to travel further, faster and more frequently in the Georgian period. By the 1770s, 1,000 hackney coaches catered for journeys within the city and this was in addition to the thousands of private carriages used by London's wealthiest residents, discussed in Chapters 5 and 6.[6] England's stagecoach system, which mushroomed in the second half of the eighteenth century, also converged on London. By 1825, there were 600 short-stages, about one-fifth of the total in England, and these made around 1,800 journeys a day to and from the suburbs; the expansion of long-stage services was equally impressive and by the mid-1770s, around fifty of the city's inns were directly involved.[7] Notwithstanding the importance of these developments, we should keep in mind that London remained a city of walkers. The majority of its residents could not afford to travel in a horse-drawn vehicle but even wealthy Londoners often opted to walk to an extent that astonished many European visitors.[8] Coach travel has become a defining image of the Georgian era but it was in the transportation of goods, even more than passengers, that draught horses made their most significant contribution in London.

The city benefited from major improvements in draught horse breeding in this period. At the start of the eighteenth century, heavy horses of manifold build, strength and colour worked on English farms, but by the 1750s, experimentation with local and imported horses had brought a handful of elite breeds to the fore. At first they were bought by farmers in neighbouring counties, but metropolitan dealers were hot on their heels. Many of the capital's waggon and cart horses were described as Midland Blacks, a loosely-defined protean 'breed'. Developed principally in Leicestershire, these animals carried the blood of sturdy Flemish imports introduced in the seventeenth and early eighteenth centuries. After two and a half years, they were sold to arable farmers in the South Midlands and the West Country, where they were broken in and introduced to moderate work in the fields. It was not until these horses reached maturity at five or six years of age that they became eligible to work in London.[9] Midland Blacks varied in size, shape and strength, and the breed continued to evolve. This helps to explain why we find these horses performing such a wide range of haulage work in the city. Geldings were usually preferred to mares because they were stronger, but also to stallions because they were found to be more tractable. In the second half of the eighteenth century, Robert Bakewell and other breeders recognised that 'strength and activity, rather than height and weight' were 'the more essential properties' for horses performing heavy haulage work. To obtain these qualities improvers used selective breeding to gradually shorten the chest, neck, back and legs of Midland Blacks while fostering 'a short thick carcase' and retaining leg thickness. The agricultural writer William Marshall noted that this 'improved variety' out-performed the 'loose heavy sluggish sorts of this breed' and was able 'to carry flesh, or stand hard work, with comparatively little provender'.[10] Contemporary prints and paintings suggest that Midland Blacks were ubiquitous in the streets of London but they probably remained an elite minority among a miscellany of lesser draught horses. Tradesmen and shopkeepers employed an array of less powerful horses, reared throughout the English countryside, to draw light town carts. Many of these animals began their careers in the hackney and stagecoach trades but were sold on when they lost fleetness of foot. Dealers habitually advertised these animals as being 'strong', 'stout', 'full of good meat', 'seasoned' and 'just out of constant work' but this patter masked huge variation in anatomy, health and behaviour.

London generated an astonishing amount of haulage work for horses. Between 1680 and 1830, the number of waggon carrying services converging on the city trebled, and by the end of the period about 1,000 of these horse-drawn juggernauts traversed the metropolis every week. This work was a lifeline to many provincial economies and benefited the entire country, but above all waggon horses served London. They brought grain, vegetables and other foods into the city; they supplied its industries with raw materials; they enlarged the hinterland of its port; they distributed its manufactures; and they supported its role as a financial centre. Some waggon firms were based in the city but many were headquartered in provincial towns and when required to rest their horses, they often used country stables to reduce costs, meaning that these animals rarely had more than a transient presence in the city.[11] Meanwhile, however, a huge amount of intra-urban haulage work was performed by cart horses constantly employed and stabled in the city. For this reason, and the fact that historians have almost entirely overlooked them, these animals deserve the undivided attention of this chapter.

Smaller and more manoeuvrable than waggons, carts were used to carry a panoply of goods and waste between wharves, warehouses, markets, shops, houses, pits and construction sites. They varied in size and design and could be pulled by anything from one to four horses. Two-wheeled carts were legally entitled to carry loads of up to one ton – this was raised by one quarter in 1757 – but heavier loads were carried, both by four-wheeled carts and by law-breakers.[12] There were three main categories of cart: those licensed for hire in the City, those privately employed by tradesmen, and suburban errand carts. Licensed carts came into being in 1512, when the Court of Aldermen granted forty freemen permission to ply for hire in the City. This figure steadily increased until, in 1654, an Act of Common Council established a limit of 420, which remained in place throughout the Georgian period. In theory, licensed carmen enjoyed the exclusive privilege of acting for hire in the Square Mile of the City. Much of their business came from the Port of London wharves, picking up freight and delivering it to warehouses. When trade was buoyant, carts overwhelmed the area around Thames Street, but all the while merchants and tradesmen were quietly investing in their own haulage operations.[13] Eventually these private carts became too numerous for the licensed carmen to ignore and in 1772 they deliberately obstructed them from collecting property at the riverside. Enraged, the owners appointed a committee which threatened to prosecute any carmen who dared to persist. These plans were sent to the *Gazetteer* newspaper and published, but not without a spirited riposte from the Fellowship of Carmen appearing immediately below. They assured the public that 'a few cheesemongers of this city and suburbs who keep carts' would not 'intimidate them', but the battle was already lost. Unable to increase their numbers, the licensed carmen were powerless to stop private carts devouring much of the new distribution work generated by industry and overseas trade.[14]

Some of the fastest-growing demand for draught horses came from the construction trades. House-building has received limited attention in histories of the Industrial Revolution because it experienced so little technological change but, as Linda Clarke has argued, this overlooks major developments in the organisation of labour, materials, equipment and finance.[15] Metropolitan expansion and improvement depended on the distribution of enormous quantities of timber, bricks, stone, slate, gravel, lime and other materials. By 1790, more than 100 timber merchants traded in London and the largest kept yards close to the Thames, where English, Scandinavian and American woods were landed at several wharves. Once loaded, four-wheeled timber carriages drawn by three or more horses, together with smaller carts, made their way to building sites peppering the metropolis.[16] Horses played an equally important role in the brick trade. To reduce haulage costs, production clustered as close as possible to building sites but back-to-back local deliveries kept cart horses busy throughout the day. The Old Bailey trial of Benjamin Hall, a brick carter accused of theft in 1809, highlights the intensity of the work performed by both men and horses in this trade. On 20 September, Hall set off from a brick-maker's yard at Whitechapel Mount with a single horse and a cart fully loaded with 500 bricks. After unloading at Wentworth Street in Spitalfields, the co-workers delivered 1,000 bricks in two outings to the Swan tavern in Bethnal Green. By then, Hall's horse had hauled more than four tons of bricks over a total distance of 11 km and this only amounted to half a day's work. In the afternoon, the pair headed to Hanbury's brewery in Brick Lane with another load. Hall claimed that on arrival he had left his horse and cart in the care of a boy, went for a beer, and when he returned the bricks had gone. Failing to convince the court, Hall was publicly whipped through Whitechapel and put to hard labour for six months in the house of correction.[17] All carters in Georgian London lived precariously and so the temptation to increase their earnings by various illicit means was hard to resist. The metropolitan press occasionally described carmen as 'impudent', 'careless' and 'wicked', but financial pressure and hostile working conditions clearly influenced their behaviour.[18] Newspaper accident reports often blamed carmen, as well as hackney coachmen, for driving 'furiously' or 'in a hurry' to complete jobs faster and reach new opportunities ahead of their competitors. In 1761, for instance, the *St James's Chronicle* reported that a boy crossing the street in Bishopsgate had been crushed between the wheels of two carts, one of which was hurrying 'to get first to a House in the neighbourhood' where he had been called.[19]

Demand for architectural stonework, particularly in the West End, generated further work for draught horses, as did the improvement and maintenance of paved streets. From the mid-eighteenth century, Commissioners of Sewers and Pavements in the City and Westminster regularly hired paviours to remove, cart away and dispose of broken stones, as well as to collect heavy materials from the river. In February 1767 alone, the City Commissioners contracted with one firm to deliver fifty tons of 'square Guernsey Pebbles for paving the Carriage Ways' and advertised for another to take up and newly pave 25,000 yards of carriageway and 10,000 yards of 'Purbec paving in the Footways'. Much of this work was self-generating because the intensification of horse-drawn traffic both damaged existing paving and encouraged the construction of additional roads.[20]

As metropolitan trade and consumption multiplied, the city's draught horses distributed an increasingly diverse and valuable range of goods. By the mid-eighteenth century, hundreds, if not thousands, of tradesmen and shopkeepers would have commanded a horse-drawn town cart. They included brewers, grocers, tea dealers, butchers, dyers, flour millers, bakers, stationers, tallow chandlers, stove-grate-makers, fellmongers, shoemakers and gardeners.[21] Topographical views of London almost invariably feature horse-drawn carts, the only exception being West End scenes, which generally concealed evidence of trade in favour of polite carriages. By contrast, *A Bird's Eye View of Covent Garden Market*, 1811 (see Figure 8)

8 Detail of John Bluck after Augustus Charles Pugin and Thomas Rowlandson, *A Bird's Eye View of Covent Garden Market* (hand-coloured aquatint, 1811).

delights in the sight of dozens of horse-drawn carts and makes it easy to imagine the cacophony of clattering hooves, whinnying, rattling wheels and sales patter. Many of those who visited Covent Garden were restocking shops but some grocers and chandlers offered home delivery by cart.[22] Draught horses enabled tradesmen to extend their reach and when firms painted their names on the side-panels of carts, they became valuable mobile marketing tools. At the same time, a horse-drawn cart gave small businesses much-needed flexibility in a seasonal and uncertain commercial environment. The Old Bailey Proceedings reveal that some Londoners used horse-drawn carts for different purposes over the course of the year. In 1765, for instance, Thomas Ibetts, a greengrocer, kept a horse and cart to collect vegetables from Covent Garden Market and 'at spare times' to dig and carry gravel. Similarly, in the 1790s, Thomas Pragnall was known to go about with a horse and cart to sell 'greens and garden stuff' in the summer and to collect dust in the winter. And in 1769, William Whitley of Wapping described himself as 'a shoemaker by trade' but confirmed that he had gone to Smithfield to buy a horse because 'I keep a little cart… to do jobs'.[23]

Owning a horse-drawn cart was well beyond the reach of most costermongers, London's street sellers. Many of these individuals were forced to carry baskets on their head and trudge through the streets, crying their wares as they went, but some worked with a jackass. These animals could be bought for as little as 5s, ten times less than a modest cart horse, but even this remained a significant expense for the labouring poor. In the 1840s, Henry Mayhew estimated that 12,000 men and about 1,700 donkeys were employed in this unforgiving line of work. Men and to a lesser extent women, of all ages, worked with jackasses but adolescent males were particularly prevalent. Generally employed by the animal's owner, these lads trudged through the capital crying fruit, greens, potatoes, fish, poultry and other foodstuffs, but also firewood, brick dust and other household wares. Panniers were often slung over the backs of jackasses but some were made to draw small carts. One of these partnerships caught the eye of Paul Sandby as he worked on his series of *London Cries*.[24] Sandby revels in the boy's insolent demeanour and plays

9 Paul Sandby, *London Cries: Boy with a Donkey* (ink and graphite on paper, *c.* 1759).

up the scruffy similarity of these scrawny co-workers (see Figure 9). It is striking that when street sellers testified at the Old Bailey they tended to draw attention to their jackasses. Presumably, this was because in the finely graded hierarchy of London's labouring poor, these animals elevated men above the poorest hawkers who walked alone. Nevertheless, an air of disrepute still clung to anyone who 'went about with jackasses' – these individuals were often accused of blasphemy, assault and theft.[25]

As draught horses facilitated increasing consumption, their work also multiplied in the collection and disposal of waste. Nightmen emptied cesspools and chimney-sweepers extracted soot, leaving all manner of other detritus for rubbish-carters and scavengers. These men provided a vital and profitable public utility; the city generated mountainous quantities of human faeces, animal dung, food scraps, cinders, rubble and industrial waste, the collection and carriage of which required specialist tools, vehicles and knowledge. For this reason, only a few operators mastered more than one service. In the 1750s, William Lewis combined being a tallow-chandler, rubbish carter and nightman; his trade card advertised that he 'Keepeth Carts and Horses for Carriages of Sand, Gravel, Slop, Rubbish, &c' as well as 'Night-Carts and Men for emptying of Bogg-Houses'. Meanwhile, similar cards engraved for the nightman and rubbish carter, Robert Stone, depicted a vehicle specially adapted for storing sewerage and a more conventional two-wheeled cart loaded with barrels.[26] Horses employed in these trades expended much of their energy travelling to rubbish pits on the outskirts of the city. One of the most impressive was at Battle Bridge (now King's Cross) which, by the 1830s, forced horses to scale a colossal heap of unstable filth.[27] When employed to remove rubbish from building sites, scavengers would often return from suburban pits loaded with gravel, screed, clay and sand to double their earnings. The same men collected the cinders swept from thousands of fireplaces and sold them to brick-makers on the edge of the city. A decent living could be made from waste removal and by the late eighteenth century some master dustmen and nightmen mustered impressive fleets of horse-drawn vehicles.[28] Mr Hands of Old Street employed seven horses with a cart for each, and the same number of men to 'collect dust about the metropolis', some of which he sold to a brick-maker in Hoxton. Horses working in these trades not only had to endure putrid stenches but also to remain calm as sewage sloshed, rubble clattered and shovels scraped, an unnatural challenge for animals equipped with such highly tuned senses, an issue to which we will return.

These were not the best-treated of London's draught horses but even basic levels of care placed heavy demands on already hard-worked men. In 1764, Thomas Legg noted that between ten and eleven o'clock on a Sunday night, 'The Slaves who do Business for Nightmen' could be seen 'preparing their Teams of Horses, to come into the City, and follow their Occupation'.[29] Purchasing, feeding and maintaining draught horses also proved costly for business owners.

A cart gelding could cost as much as £12 in the mid-eighteenth century, a major investment for any small business; and once purchased, these animals proved a constant drain on resources. Some tradesmen and shopkeepers would have shied from the prospect of receiving regular bills for provender, stabling, tackle, shoes and farriery, opting instead for wheelbarrows, jackasses and hired porters to deliver light loads more affordably. Yet, these were no replacement for a horse-drawn cart and for this reason many businesses began to hire errand carts in the second half of the eighteenth century. The earliest reference to such a service appears in a newspaper advert from May 1774, which lists access to daily stagecoach and errand cart services among the benefits of a 'neat small cottage' to be let in the village of East Ham, 10 km from Whitechapel. An errand cart first appears in the Old Bailey Proceedings in January 1780, when items of clothing were stolen from Richardson's Tottenham-bound vehicle; and they appear to have become much more prevalent in the 1790s, a decade in which poor harvests made it far more expensive to maintain a horse.[30] Between 1780 and 1820, the Proceedings record seventeen errand cart services including Hampstead and Tottenham in the north; Blackwall and Poplar in the east; Chelsea and Dulwich in the south; and Hammersmith and Richmond in the west. Errand carts dispatched a vast array of goods for clients as diverse as cheesemongers, linen-drapers and glass merchants. In January 1800, a tobacconist in Bishopsgate used the Woolwich errand cart to deliver a parcel of tobacco, worth 15s, to a customer at the King's Head in Woolwich; and in 1808, John Turner, a City warehouseman hired the Lambeth and Vauxhall errand carter to carry a packet of stolen indigo, valued at £20, to a house in Paradise Row.[31] The rise of the errand cart illustrates that draught horses simultaneously served London's industrial and consumer revolutions; but it was in the distribution of two commodities, in particular, that these animals made their greatest contribution to the metropolitan economy: coal and beer.

By the 1640s, ships were already bringing 150,000 tons of coal into London every year; and by 1800, this had rocketed to 1.2 million tons, which equated to approximately twelve million sacks or 400,000 waggon-loads. London's voracious consumption of this bulky fuel generated copious work for horses, requiring them to transfer cargo from the wharf to the warehouse and on to customers. Increasingly, this work was fulfilled not by licensed carmen but by independent coal merchants whose assets included large stables, a fleet of vehicles and several servants to drive them. Rowland de Paiba, a coal merchant of Upper Thames Street in the early 1770s, owned twenty-four horses, nine coal carts, one waggon and seventeen barges and lighters, placing him in the upper echelons of the trade. In the 1840s, Henry Mayhew estimated that a large wharf employed thirty horses, six waggons and four carts, with seven men to drive them, and fifteen to twenty coal-porters. Meanwhile, a small wharf required ten horses, three waggons, one cart, four drivers and five porters. By 1790, the *Universal British Directory* listed 203 coal merchants in London, which would suggest that the trade already employed at least 2,000 horses.[32]

The most valuable livestock belonged to an elite group of around thirty wholesale merchants, or first buyers. Once fully loaded, their coal waggons required four strong horses and hauling several tons uphill from the Thames put immense strain on their bodies. When rain, mud or ice coated the stones, slips could lead to spine-chilling accidents and crippling injuries. The majority of coal sellers operated on a much smaller scale and delivered direct to households using more nimble carts drawn by one or two horses. On arrival, the horses were trained to stand still for several minutes as their co-workers unloaded. But coal horses did more than supply homes, they also fed the furnaces of numerous heat-dependent industries including brewing, soap-making, sugar-refining and glass-making. In 1805, Briant and Back made regular waggon deliveries from their wharf at Wapping to Hanbury's brewery in Spitalfields.[33] Two years later, the brewery's coal needs soared when Hanbury installed the firm's first steam engine, meaning that technological innovation promoted rather than threatened traditional horse work.

Distance, however, did pose a serious challenge for coal dealers and their horses. In 1771, Thomas Cranage was obliged to send three carts on a 13 km round trip from the Thames to Kentish Town, then a satellite village to the north of the city. Urban sprawl forced dealers to adapt their cartage rates: in the 1790s, for example, J. Williams of Wapping advertised that he delivered 'to any part of the Town' at 4d per chaldron and to the 'Country at 1 Shilling per mile extra off the Stones'. Much of these cartage fees would have been spent on fuelling and maintaining London's largest and strongest draught horses. By the late nineteenth century, an average coal horse distributed thirty tons of coal a day, working eighty hours over a six-day week.[34] Only the finest Midland Blacks could withstand this level of work and coal merchants advertised this with great pride. From the mid-eighteenth century, horse-drawn carts take centre stage in the majority of surviving trade cards commissioned by coal merchants. An early example, engraved for Philip Fruchard of Thames Street, depicts a horse-drawn cart being loaded with coal (see Figure 10). Fruchard's horses wait patiently for their instructions as his workmen shovel coal into sacks and carry them across a gangplank. An incoming ship places Fruchard's horses at the heart of a nationwide network of men, horses and machines dedicated to extracting and distributing Britain's 'black diamonds'. In the coal fields of northern England, Galloway pit ponies dragged sledges along narrow shafts guided by similarly compact human mineworkers, often children. London demanded a different kind of co-operation but one that still led men to question which species had the rawest deal.[35] In the 1840s, Henry Mayhew quoted a seasoned coal-backer as saying:

> Our work's very hard… First, you see, we bring the coal up from the ship's hold. There sometimes it's dreadful hot, not a mouthful of air, and the coal-dust sometimes as thick as a fog… Then there's the coals on your back to be carried up a nasty ladder… perhaps 20 feet-and a sack full of coals weighs 2 cwt. and a stone at least; the sack itself's heavy and thick. Isn't that a strain on a man?

10 Anon., draft trade card of Philip Fruchard, coal merchant (etching, *c.* 1750).

Answering his own question, the coal-backer complained 'No horse could stand it long.'[36] Increasingly, however, coal dealers issued trade cards in which horses overshadowed men. In the 1780s, for instance, Benjamin Levy demonstrated his firm's prowess by foregrounding a pair of horses effortlessly hauling a heavy load up the steep banks of the Thames. In doing so, Levy relegated his carter and two other workmen to the edge of the scene where they appear as little more than spectators.[37]

The Georgian period laid the foundations for even more extensive employment of coal horses in Victorian London. Population growth, industrial expansion and, from the 1840s, London's emergence as a railway hub increased the city's annual consumption of coal from 1.25 million tons in 1810 to six million in the 1860s and eight million in the 1890s. Mayhew estimated that by the 1840s, more than 5,000 horses were drawing 1,600 coal waggons and 400 carts through the city. Fifty years later, William Gordon claimed that the number of horses employed had risen to 8,000.[38] The coal trade probably overtook brewing as London's leading employer of draught horses before 1800 but no animal embodied the city's economic prowess more than the brewer's dray horse.

A brewer producing 100,000 barrels a year required about sixty dray horses, rising to ninety for 200,000 barrels and 150 for 300,000. On this basis, the city's eight leading breweries must have employed more than 300 animals in 1780 and well over 600 in 1825. But there were never fewer than 140 breweries in this period and while most operated on a much smaller scale, by 1800 the trade almost certainly employed more than 1,000 dray horses.[39] These impressive animals attracted considerable attention, not least from artists. Horses abound in the art of Georgian England but historians tend to focus on depictions of aristocratic thoroughbreds, overlooking a thriving artistic genre which celebrated heavy working horses including the London dray horse. Around 1790, George Garrard, a painter admired by Sir Joshua Reynolds, completed a dramatic depiction of his city's premier brewery, the White Hart on Chiswell Street. The owners, Samuel Whitbread I and II, were Garrard's chief patrons and although they did not commission this scene, it clearly reflects their tastes, interests and values. The original painting has disappeared but it was reproduced in 1792 as an aquatint engraving (see Figure 11). Garrard depicts a hive of industriousness framed by warehouses from which tall chimneys belch thick black smoke. Significantly, however, the scene is not dominated by men or machinery but a gigantic horse. Appearing to combine great strength with an intelligent understanding of its role, the animal steadily backs into the shafts of a dray. Its coat gleams like a Stubbsian thoroughbred and its white legs brilliantly contrast their sooty surroundings. Garrard's choice of pose closely resembles that of Stubbs' *Horse Frightened by a Lion*, first exhibited in 1763, but here the similarity ends; this dray horse is a paragon of business-like calm, the ideal mascot for his patrons. Samuels I and II shared a passionate interest in dray horses with the artist, who completed at least two other paintings in which their animals take centre stage: *Loading the Drays* (1793) and *Mr Whitbread's Wharf* (1796), as well as a fine plaster model of a dray horse (1796).[40]

Garrard was not the first to depict dray horses but he did particular justice to their elevated status in London life and established a trend which continued well into the nineteenth century. In 1805 and

11 William Ward after George Garrard, *A View from the East-End of the Brewery Chiswell Street* (mezzotint, 1792).

1807, Dean Wolstenholme the elder foregrounded dray horses in bustling depictions of the Red Lion brewhouse in East Smithfield and the Golden Lane Genuine Brewery in Cripplegate. Fifteen years later, his son followed suit, celebrating their work at the Hour Glass in Upper Thames Street and the Black Eagle in Spitalfields.[41] Named dray horses even inspired admiring portraits by artists acclaimed for portraits of racehorses and aristocratic hunting scenes. In 1798, the celebrated animal painter, John Nost Sartorius, painted a veteran dray horse known to its owners as Old Brown. Purportedly thirty-five years old, Old Brown was an exceptionally aged horse for the period, and dubiously long-lived for a hard-worked dray horse. The apparently healthy animal rests in a tranquil corner of the brewer's yard, unharnessed, and it is conceivable that Old Brown was permitted to serve out its twilight years as a mascot, although this would have been highly unusual.[42] A later work by John Christian

Zeitter captures the power and prestige of dray horses in their prime: the heroically named Pirate and Outlaw belonged to Reid's Griffin brewery in Clerkenwell and are depicted with their tails swishing and muscles bulging as they prepare to haul barrels up from a cellar.[43] Brimming with self-confidence, one of the horses boldly turns its head to meet the viewer's gaze. In the twenty-first century, images focusing on coal, steam and iron, such as Joseph Wright of Derby's *An Iron Forge* (1772) and Philip James de Loutherbourg's *Coalbrookdale by Night* (1801), tend to be considered the most iconic images of the Industrial Revolution; but for Georgian Londoners, nothing conveyed the nation's industrial progress and prosperity more compellingly than the city's dray horses.

London's leading brewers scoured country horse fairs and employed dealers to secure the largest and strongest Midland Blacks, prices for which increased from about £16 in the middle of the century to around £40 after 1800, more than many carriage horses fetched. The qualities esteemed in a dray horse were no less specific than those sought in a thoroughbred racehorse. The vet and livestock writer William Youatt observed that a brewer's horse should be at least 17 feet in height but more importantly, they must have a 'broad breast, and thick and upright shoulders… deep and round barrel, loins broad and high, ample quarters, thick fore-arms and thighs'. These were the mighty animals which astonished newcomers to London well into the nineteenth century; they were a talisman for modern brewing and as such, proved valuable marketing tools, but size and bulk were not bred into these horses for mere show. Midland Blacks were generally deemed too slow and heavy for ploughing or drawing carts but were ideally suited to heavy haulage work in the capital. Brewers expected them to haul a vehicle loaded with three 108 gallon butts of beer, collectively weighing 1.5 tons, for twelve hours a day or more. The intensity of the work, the unforgiving nature of the vehicle and challenging metropolitan conditions demanded a special kind of strength and durability. Youatt noted that

> over the badly-paved streets of the metropolis, and with the immense loads… great bulk and weight are necessary to stand the inevitable

> battering and shaking. Weight must be opposed by weight, or the horse would sometimes be quite thrown off his legs.[44]

Dray horses were, however, admired as much for their temperament as their physical attributes. In the second half of the eighteenth century, livestock improvers repeatedly selected the most 'gentle' or manageable Midland Blacks for breeding. This served metropolitan brewers well because the city's streets were becoming an increasingly hostile environment for horses to work in. It was for this reason that William Wordsworth, appalled by the hubbub of the metropolis, expressed so much admiration for 'the sturdy Drayman's Team, / Ascending from some Alley of the Thames / And striking right across the crowded Thames / Til the fore Horse veer round with punctual skill'.[45]

Remaining calm and fulfilling instructions while ignoring a riot of distractions – including moving vehicles, barking dogs, noxious fumes and a cacophony of human voices – demanded immense self-discipline from these naturally hyper-sensitive animals. Such stimuli provoked what modern veterinary ethologists describe as 'evasions': bolting, shying and sudden swerving of the forequarters; as well as 'hard-wired agonistic anti-predator responses' such as rearing or bucking, lowering the head and kicking out with the hind legs.[46] While experienced men could sometimes manage startled animals, this was not always possible. In 1762, for instance, the horses drawing a light cart through the Kingsland Road 'took fright at the tilt of the Peterborough waggon passing by' and in the ensuing panic, the driver of the cart was thrown against the shafts of the vehicle and 'killed on the spot'.[47] Two years later, the fore-horse of a cart 'took fright at something being hastily thrown out of a house' in Shoreditch and the wheels of the cart crushed a child to death.[48] The animal painter and keen equestrian Dean Wolstenholme the elder, paid particular attention to the ways in which horses reacted to urban stimuli, as well as the measures which draymen took to prevent accidents. In his 1807 engraving, *A Correct View of the Golden Lane Genuine Brewery* (see Figure 12), a wild-eyed dray horse lurches backwards as a hogshead erupts, a cooper uses a hammer and workmen noisily roll barrels across a

12 Detail of William S. Barnard after Dean Wolstenholme, *A Correct View of the Golden Lane Genuine Brewery* (mezzotint, 1807).

yard; chaos is only averted because one of the draymen is walking alongside the leading horse. This practice had been a legal requirement since 1715 but in the 1740s–60s, draymen were heavily criticised in the London press for illegally riding on the shafts of their vehicles, from where they struggled to bring their animals to a halt.[49] In 1748, the *Old England* reported that a drayman riding on the shafts had run over two children playing in the street, one wheel breaking the boy's leg, the other leaving a girl 'for dead'. Such incidents almost certainly informed William Hogarth's spine-chilling depiction, in *Second Stage of Cruelty* (1751), of a dray about to crush a boy playing with a hoop (see Figure 2). The artist's inclusion of this tragic vignette was part of concerted campaign led by Westminster's chief magistrate, Henry Fielding, and supported by the London press, to police a new law against dangerous driving.[50] Yet, as we will see, while some drovers may have acted irresponsibly, these accidents also highlight the growing challenges that men and horses faced on the streets of the metropolis.

Once outside a victualler's premises, draymen would remove one horse from the shafts and attach ropes to its harness to raise up empties from the cellar and then lower their heavy replacements.[51] According to an 1846 issue of the *Sporting Magazine*, a gentleman visiting the city for the first time had once marvelled that these animals

> performed this office without any signal... not a word or sign escaped the man at the top of the hole, who only waited to perform his part as methodically as his four-footed mate did his... The cellaring finished, the horse took his place by the team... The man... then walked away; the team followed. Not one word had passed, not even a motion of the whip, or any other intimation of what was to be done next.[52]

A similar scene attracted the attention of the artist George Scharf the elder in the 1820s and his necessarily brisk sketches demonstrate that well-trained horses freed draymen to arrange their hooks, ropes and ramps as well as deal with victuallers (see Figure 13). This illustrates that equine intelligence made a major contribution to commercial efficiency in brewing, London's leading industry. In the 1990s, animal behaviour research confirmed what people who work with horses have long recognised: they have acute memories for patterns, routes and work routines, meaning that they can perform familiar tasks with little or no supervision.[53] But draymen and dray horses also had to develop a mutual understanding and trust. Novice horses had to be eased into their work before taking on full duties and draymen had to get to know their behavioural idiosyncrasies and acclimatise them to unfamiliar sights and sounds, as well as precipitous slopes, slippery conditions and heavy traffic. At the same time, the horses grew to trust their human co-workers and anticipate their demands.

Brewery haulage work increased dramatically in the Georgian period. In the 1690s, Londoners had fetched most of their beer in person, meaning that brewers only needed to employ a few horses but by the 1760s, the situation had changed dramatically. In a departure closely linked to the rise of the 'tied trade' – which saw elite brewers amassing the leases of

13 George Scharf, *Draymen and Horse* (drawing, *c.* 1820–30).

numerous public houses and thereby centralising the business of production, distribution and retail – breweries began delivering monthly to hundreds of publicans scattered across the metropolis.[54] The Whitbread records reveal that horses setting out from Chiswell Street in 1800 visited almost 400 victuallers peppering the metropolis from Paddington in the West (6.5 km away) to Woolwich in the East (12 km), and from Finchley in the North (13 km) to Peckham in the South (5.5 km). Two-thirds of customers were located more than 1.5 km from the brewery. The fact that dray teams could deliver to suburban pubs and return the same day was a major logistical advantage for the trade, but distribution on this scale meant gruelling shifts for men and horses alike. In 1764, a brewer in Hackney confirmed that his stables were never locked up 'because we are fetching the horses out almost all hours of the night'. In 1777, a young brewer in Wapping 'Set the drays off' in the morning and only counted them back in at nine o'clock at night. And in 1810, Louis Simond, a merchant visiting from New York, was amazed that the horses serving Barclay Perkins in Southwark were 'often sixteen hours in harness'. By then, a London dray horse could be expected to haul 2,000 barrels a year, with a combined weight of 1,000 tons.[55] This raises an important question: were dray horses worked harder as the Industrial Revolution gathered pace in the late Georgian period? A comparison of the Black Eagle brewery's production figures and livestock data would certainly suggest so. In the 1770s, the firm's dray horses each hauled around 1,433 barrels per year but by the 1820s, they managed 2,605, an increase of 82 per cent. This becomes even more impressive when we compare it to changes in human work patterns. Hans-Joachim Voth has estimated that the annual hours worked by Londoners increased from 2,288 in the 1750s to 3,366 in the 1830s, a rise of 47 per cent.[56] We cannot accurately compare how hard people and horses worked using one dataset based on tasks completed and another on hours worked, but it does appear that dray horses were harder hit by industrialisation than the average working Londoner. It is tempting to read this as evidence of worsening treatment but selective breeding transformed draught horses in the Georgian period and a dray horse was capable of considerably heavier work in the 1820s than it had been sixty years earlier. Moreover, over the same period, these animals became better fed, stabled and cared for.

Brewers became ambitious builders in the eighteenth century, partly to accommodate immense new vats and valuable steam engines, but also expanding ranks of dray horses. At the Anchor in Southwark, London's leading brewery in the early 1770s, Henry Thrale invested heavily in plant, machinery and architecture. By 1774, he had built individual stall stabling for seventy dray horses arranged in two stable blocks; the largest formed a horseshoe around a large dung pit, with forty-five stalls for dray horses and an adjoining wing for mill horses. In the north-east corner of the brewery yard, Thrale kept an infirmary which could tend to four horses at a time.[57] The year 1780 proved turbulent for the Anchor: war with the American colonies was stifling trade; in February, Thrale suffered a debilitating stroke; and in June, a mob attacked the brewery during the Gordon Riots. Nevertheless, the firm decided to press ahead with major improvements to its stables, a remarkable testament to its prioritisation of equine needs.[58] This marked a step change in metropolitan brewery stabling. By the late 1790s, Whitbread's brewery at Chiswell Street had built three stable blocks, the largest of which, the 'Great Stable', was almost 40 m long and contained separate stalls for eighty dray horses.[59] The separation of horses with wooden partitions began in the Tudor period but only in aristocratic stables. Prior to this, horses were separated by a bale or pole suspended from the roof or the facing wall. This system endured in some commercial and military stables throughout the eighteenth century but London brewers wanted to ensure that their precious livestock could rest undisturbed and safe from injuries inflicted by jostling and kicking.

The city's leading firms made use of new materials and construction techniques but also new ways of thinking about equine care and industrial efficiency. In particular, brewers and their architects developed a growing appreciation of the importance of space, light, ventilation and cleanliness. By the 1830s, Truman's stalls were 3 m in length. This afforded a modest 60 cm of clearance within the stall but each one opened onto a generous central passageway (3 m wide and 21 m long), which gave the horses plenty of space to turn in without twisting their legs or colliding with other animals.[60] When the Whig politician Thomas Creevey visited Whitbread's stables in 1823, he was impressed to find a stable 'brilliantly illuminated, containing ninety horses worth 50 or 60 guineas apiece'.[61] Light was crucial to improving equine care, enabling stable workers to identify injuries and early signs of disease, as well as helping them to keep the stable clean and move animals safely. Further improvements were to follow: in 1837, Truman's Black Eagle in Brick Lane unveiled a stable for 114 horses, which cost almost £10,000, around the same as Lambeth's Church of St Andrew. *The Civil Engineer and Architect's Journal* noted that the four stable blocks contained 300 tons of iron, including an innovative cast iron frame, columns, brackets, cantilevers, drains, troughs and mangers. Above the iron manger in each stall, a cast iron tablet recorded the occupant's name to help ensure that each animal received individualised care. A key feature of Robert Davison's design was its attention to sanitation. Iron grating in the centre of each stall fed waste into a cast iron drain which ran the entire length of the stables and which was flushed by a stream of water for a few minutes every morning. As well as benefiting from hot and cold water, the stables were ventilated from above and below by air-bricks, flutes and 'moveable ventilators to regulate the egress of foul air'. According to the *Civil Engineer*, the stable's many achievements included reducing the risk of disease, the prevention of splinters injuring the horses, security against fire, and the durability of its cast iron fittings, which stood up to 'the rough usage they are subject to by the dray-horses'.[62] Such sites were engineered to prepare dray horses for increasingly hard work and to maximise their productivity, but this also required improvements in nutrition and care.

When Louis Simond visited Barclay Perkins' dray horses in the 1810s, he noted that 'These colossuses are fed with a mixture of clover-hay, straw, and oats'.[63] Simond recognised that these hard-worked animals demanded an invigorating diet and by 1840s, the average daily calorific consumption of a dray horse at work was around 38,000 kcal.[64] By comparison, the average horse at rest only consumes 12,000–15,000 kcal per day.[65] We know that at least some brewers improved the nutritional regimes of their horses during the Industrial Revolution because the Truman brewery rest books record what its animals ate from the 1770s into the nineteenth century. Between 1771 and 1775, clerks recorded the annual quantities of provender purchased for the brewery's horses; thereafter, we are only told the quantity of foods on-site at the time of the summer stock-take. This makes it impossible to track changes in the quantity of foods fed to dray horses but the records do reveal that a crucial dietary change was made in the late eighteenth century. Throughout the 1770s, the animals were fed on hay, chaff (a mixture of cut meadow hay, straw and clover), wheat bran and pollard (reject flour). By 1791, however, when the surviving records resume, three nutritionally rich foodstuffs had been added to the menu: beans, oats and clover.[66] The nutritional benefits of these foods were well known by the late 1790s, when John Middleton published his *View of the Agriculture of Middlesex* (1798); and in the 1830s, William Youatt confirmed that oats afforded 'the principal nourishment' for horses and that beans, given principally in winter, added 'materially to the vigor [sic] of the horse', without which 'many… will not stand hard work'. Youatt advised that a mixture of 'eight pounds of oats and two of beans should be added to every twenty pounds of chaff'.[67] Clover, an energy- and protein-rich legume, was another valuable addition to the dray horse's diet, and was given in the late spring and summer as a supplement to hay. The systematic introduction of beans, oats and clover in the late eighteenth century was scientifically calculated to increase strength and stamina in order to prepare dray horses for harder work; and it seems likely that all of London's leading brewers, and at least some smaller firms, followed suit. It is important to note that these foods were relatively expensive items on which to feed horses and that brewers introduced them at a time when their prices were still rising.[68] In 1795, a petition signed by twenty-two metropolitan hostlers complained that 'the heavy prices of Horse Provender' which they had 'borne for some time' and which were 'daily increasing… render it impossible… to do justice to the Horses under their care, unless they are permitted to make an additional charge' to their customers.[69] But London's brewers calculated that a well-fuelled equine workforce would more than offset these growing costs.

In addition to feeding their horses better, these businesses provided increasingly sophisticated care to maximise their output and working lives. Major firms employed large teams of dedicated horse keepers and draymen. In the 1790s, a third of Barclay Perkins' workforce dealt directly with horses, a ratio likely to have been matched by competitors.[70] Alexander McLeay, a wine merchant in the city, described the painstaking work involved; the stables, he observed,

> are daily cleaned and littered, by men whose business it is to attend them night and day… they pay the greatest attention to their being and they are well littered at all seasons… They are… well rubbed down after they come in from work.[71]

Particular care was taken to protect the hooves of dray horses, which suffered heavy wear on the streets of the metropolis. This necessitated a continuous programme of regular horseshoe removal and replacement;

in the first quarter of the nineteenth century, Truman's horses consumed around 1,200 shoes and 10,000 nails every year.[72] Brewers appear to have overseen sophisticated stable regimes supported by meticulous records. By the early nineteenth century, most large firms would have kept a stable book of the kind which survives for Barclay Perkins for 1827–39. Here, the company recorded weekly tallies of sick or lame animals, deaths, total stock and additional notes. In an average week in 1828, the brewery employed 126 horses, of which six had to be rested for reasons of sickness or lameness. When Louis Simond visited a few years earlier he was impressed to find that none of these horses were sick, which he took as evidence of the superior care they received; but with so many animals stabled in close proximity, infectious diseases posed a grave threat. In a single week in December 1827, Barclay Perkins workmen found three horses dead.[73] With so much capital invested in these animals and income so reliant on their work, brewers made huge efforts to prevent sickness and injury. The 1774 plan of the Anchor brewery shows that infected animals were kept away from the main stable complex and that large gates were used to help prevent the spread of disease.[74] Trade ledgers also reveal that Whitbread spent large sums on medicine and farriery equipment and by the 1790s the firm maintained two large 'farrier's shops' on-site. By the early 1840s the same brewery employed its own veterinary surgeon in an on-site laboratory, as well as commanding a blacksmith's shop and a harness-maker's shop.[75]

As significant as these developments were, however, it remained the case that a dray horse entering a brewer's service at five years of age was generally worn out by the time it was twelve. By contrast, many other draught horses in London survived five years longer and some farm horses lived to more than twenty.[76] Metropolitan breweries extracted as much labour from dray horses for as many years as they could supply it, before selling them on, dead or alive. Brewery horse care was advanced for the period and while comparable developments can be found in other sectors – including waggon and stagecoach operations, the coal trade and waterworks – many London draught horses continued to be under- and improperly fed, overworked, poorly stabled, beaten and denied adequate care by employers who were ignorant of their animals' needs, callously unwilling to satisfy them or too poor to do so.[77] Nevertheless, in leading metropolitan industries like brewing, the intensification of equine labour could go hand-in-hand with improvements in care, mirroring the experiences of many working people in this period. Across the country, many farm and factory workers discovered that higher pay and improved standards of living accompanied longer shifts and harsher conditions. Historians continue to disagree over whether the Industrial Revolution improved or blighted human lives and the impact on horses is no less complex.[78] As we have seen, the lives of horses were entangled with those of their co-workers. Being in command of horse-drawn vehicles gave plebeian men considerable power and there is plenty of evidence in newspaper reports and court records of draymen, carters and waggoners, as well as hackney coachmen, subverting the 'deferential choreography' of metropolitan streets by jostling for position, refusing to give way and obstructing other road users.[79] Disconcerted by this, polite society complained that these men behaved like brutes and implied that this was due to spending too much time in the company of animals. Whether this was fair or not, what is clear is that increasingly heavy workloads and traffic made the work of both men and horses more challenging, exhausting and dangerous in this period. Horses took fright at urban stimuli and were often injured in accidents but their human co-workers also suffered. While the metropolitan press claimed that men rode on the shafts of their vehicles because they were lazy, this behaviour may have stemmed from genuine physical exhaustion.[80] In some situations, it also appears that men had to leap onto their vehicles to avoid serious injury: in 1736, a carter pleaded at the Old Bailey that he had been forced to do so 'or I should have been squeezed' by an oncoming dray.[81] He was acquitted of manslaughter but the drivers of horse-drawn carts and waggons were some of the most strictly regulated of London's workers. In 1715 and 1750, Parliament legislated to punish riding on the shafts with fines; and in 1757, it was ordered that all vehicles bare the name of their owner and an identification number, which had to be registered with the Commissioners of the Hackney Coach Office.[82] Draymen and carters were often arrested and charged, but the law struggled to keep pace with the unprecedented proliferation of commercial traffic in the late eighteenth and nineteenth centuries.

It is important to emphasise that brewing, like many other metropolitan trades, became increasingly reliant on horses in the nineteenth century as the use of steam power exploded. Between 1810, when Hanbury sold his remaining mill horses, and 1835, the firm's dray horse stock almost doubled from 57 to 103, and this was far from unusual. Barclay Perkins installed its first engine in 1786 followed by another, three times its power in 1832. The firm's stable book shows that this upgrade had a dramatic impact on haulage operations. In the five years prior to installation, the firm's average dray horse stock had been 113. In the subsequent five years, that figure rose by more than a quarter to 143. Then, in the first quarter of 1839, an additional thirty horses arrived, a sudden increase of more than a fifth.[83] Without dray horses, the advent of mass production brought about by steam in the early nineteenth century would have been futile. Trains were of little use to brewers as a means of distribution because the vast majority of their customers were London pubs and the only way to supply them was by road. Only a few elite brewers ran 'country' trades, for which trains were employed, but this remained a tiny component of their business. As a result, brewers had to ensure that their draught horse numbers kept up with production.[84] Engravings and early photographs show that dray horses continued to dominate the public spectacle of brewing throughout the nineteenth century and far from symbolising a romantic bygone age, these animals advertised a flourishing modernity. London's leading breweries first became curiosities in the late eighteenth century and visitor accounts leave no doubt that dray horses impressed as much as steam engines. In May 1787, *The London Chronicle* reported that George III had visited Whitbread's brewery. He spent half an hour examining the steam engine which, the paper noted

> has saved much animal labour. But there remains much labour that cannot be saved. This particularly impressed the King; for he saw... 80 horses all in their places... [and] accurately guessed the height of [one]... which was really remarkable, no less than 17 hands three inches.[85]

A decade later, a visit to Richard Meux's brewery in Clerkenwell 'amazed' the Swiss scientist Marc-Auguste Pictet, on account of '*une machine à vapeurs, qui égale en force 28 chevaux*', but also the firm's '*cinquante-huit superbes chevaux, du prix de 50 liv[res] Sterl[ing]... [qui] sont occupés à charier la liqueur dans Londres et ses environs*'.[86] And in the early 1830s, the Black Eagle hosted a group of politicians led by the prime minister, Lord Grey, which, after perusing the engine and taking dinner, went 'to the stables to see the horses'. Here, the Lord Chancellor, Henry Brougham, 'selected one of the best of them, and pointed out his merits'.[87]

From chancellors to costermongers, Georgian Londoners were acutely aware of the importance of horse work. By contrast, innovation-centric habits have led most historians to overlook the contribution made by men and horses working together. This helps to explain why London's central role in the Industrial Revolution has been so misconceived and undervalued, but the work performed by mill and draught horses is only part of the picture. In Georgian London, animals were also boom commodities in their own right and played an important role in Britain's consumer revolution.

3

Animal husbandry

Late one afternoon in July 1790, a tailor's assistant walked into Tothill Fields in Westminster, seized a sow and, it was later alleged, attempted to commit bestiality. When the animal escaped, Mathew Mulvie made a second 'attempt' with a cow and caught breeches-down by the cow-keeper's son, who was watching from a haystack, the young man was arrested. At the Old Bailey, Mulvie admitted that he had been 'very drunk' but insisted that it would have been 'madness to perpetrate such a crime in such a publick place'. The jury decided that the sexual act 'had not been completed' and after a brief detention 'for the misdemeanor', he was freed.[1] Mulvie's behaviour and the attitudes of his contemporaries towards bestiality are intriguing but far more important in the context of this book is what this case reveals about animal husbandry in the city. Mulvie's misadventure reveals that livestock were being farmed at the very heart of the late eighteenth-century metropolis, in a built-up area just a few hundred yards from Westminster Abbey, Parliament, Buckingham House and the West End. Moreover, the fact that Mulvie worked in tailoring indicates that London's agricultural activity brought a broad spectrum of people into close contact with livestock, even when they had no direct involvement in associated trades.

Visitors to Georgian London soon discovered that they were only ever a few hundred yards from grazing cows. Thomas Bowles' engraving, *The North Prospect of London taken from the Bowling Green at Islington* (*c.* 1740), depicts the capital as an island in a sea of pasture (see Figure 14)

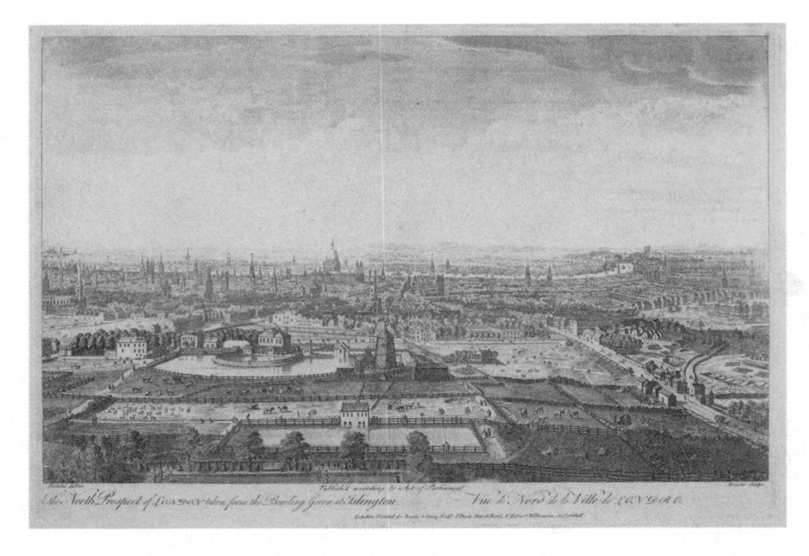

14 Detail of Thomas Bowles III, *The North Prospect of London taken from the Bowling Green at Islington* (engraving, *c.* 1740).

but some Georgian commentators exaggerated how large and urbanised London was, either to celebrate or condemn the changes they were seeing. In 1738, for instance, one topographer proclaimed 'We may call it [Middlesex] almost all London, being inhabited chiefly by the Citizens'. And historians have reinforced this view – W. K. Jordan asserted that by 1600 'London was Middlesex' – apparently overlooking the rusticity of London's immediate surroundings, and the fact that throughout the eighteenth century, Middlesex remained a predominantly agricultural county.[2] The role played by farming in metropolitan life has been similarly downplayed. Charles Phythian-Adams described London's emergence as a modern city through the juxtaposition of soot – symbolising the city's urbanisation and industrialisation in the eighteenth century – and milk, representing an earlier epoch when the city was supposedly more in touch with rural life. Yet, this dichotomy is too simplistic; just as wooden buildings now labelled vernacular were common in many parts of Georgian London and not considered rural, livestock-keeping was part of an urbanity strikingly different from that which we recognise in the twenty-first century.[3] At the same time, historians usually associate the improvement of animal husbandry in this period with farmers and landowners far removed from London.[4] Metropolitan husbandry only receives attention in studies of nuisance and poverty, which give the impression that cow- and pig-keeping could only have occurred on a small scale in filthy backyards concentrated in the city's poorest districts, and that rearing livestock was always incongruous with urban life.[5]

It has also been widely assumed that urbanisation and industrialisation in the Georgian period brought about a growing separation between sites of food production and consumption in Britain, particularly in the capital, and that Londoners were overwhelmingly reliant on the countryside for their milk and meat.[6] The geographer Peter Atkins has already challenged this view, pointing out that London was producing huge quantities of milk as late as the 1890s, at which time 15,000 acres of land within the County of London, one-fifth of its total area, were still in agricultural use.[7] This chapter does not seek to pin down a chronology of separation but to argue that animal husbandry flourished at the interface of tradition and change in Georgian London. For John Berger, the fact that 'A peasant becomes fond of his pig and is glad to salt away its pork' evinced the 'existential dualism' which, until the modern age, had always underpinned man's relationship with livestock. What the modern 'urban stranger' finds so difficult to understand, he argued, 'is that the two statements in that sentence are connected by an *and* not by a *but*'.[8] It might not be possible to identify when this bond was severed, but this chapter will present evidence which shows that it permeated metropolitan society well into the nineteenth century. In doing so, it will demonstrate that animal husbandry was a thriving sector of the metropolitan economy which adapted to urbanisation and industrialisation in remarkable ways.

By the mid-eighteenth century, English dairy farmers had superseded the Dutch and, as shown by an anonymous caricature published in 1778, the nation's milk became imbued with patriotic meaning.[9] *A Picturesque View of the State of the Nation* depicts an American congressman sawing off the horns of an English milch cow which, as the *Westminster Magazine* pointedly explained, 'are her natural strength and defence'. A jealous Dutchman milks the animal 'with great glee' while French and Spanish rogues carry away full bowls. Rather than defending the cow, the British lion dozes, mirroring events across the Atlantic, where the nation begins to lose its grip on its American colonies.[10] The use of a milch cow to symbolise prosperity would have struck a particular chord in London because it was here that the demand for, and production and marketing of, milk were at their most advanced. Operating in parallel with the gargantuan trade in meat on the hoof, discussed in Chapter 4, cow-keeping was part of a sophisticated livestock economy which made London an agropolis. It is unclear how many cows were involved; in 1794, a Soho-based land surveyor, Peter Foot, gave the new Board of Agriculture the figure of 8,500 animals, but we should treat this with caution. Around the same time, the agricultural writer Thomas Baird estimated that approximately 9.8 million gallons of milk were sold in the city every year, a trade worth half a million pounds.[11] Cow-keeping brought some families significant wealth: in 1773, the *Gazetteer* announced the sale of a business in Park Lane which owned 'sixty young milch cows, two young bulls, twelve stout able geldings of the draught kind' plus several carts and a waggon. Operations of this scale were not unusual and newspapers often referred to cow-keepers as being wealthy: in 1736, for instance, the *London Daily Post* reported that a 'great cowkeeper', Mr Capper of Tottenham Court Road, was 'said to have died worth 30,000l'.[12]

Mapping the city's cow-keeping activity is challenging but drawing on evidence from newspapers, Old Bailey Proceedings, insurance policy documents, Middlesex Sessions papers, wills, Westminster poll books and other records, I have managed to locate 258 cow-keepers active between 1730 and 1800, and to unpick the trade's development in this period.[13] Throughout the eighteenth century, London's dairy herds were concentrated in an arc fringing the northern limits of the city, from Marylebone and St Pancras in the west to Islington, Clerkenwell, Bethnal Green, Hackney and Shoreditch in the east. This area provided rich pasture enabling the city's cow-keepers to maximise milk yields while remaining close to urban consumers. Around half of the cow-keepers identified belonged to this area but there were significant concentrations of activity elsewhere. Milk remained a semi-luxury throughout the Georgian period and demand for freshness drew cow-keepers towards the increasingly populous and wealthy West End. For this reason, Tothill Fields, a thirty-five-acre pocket of grassland between Westminster and the Thames, served at least two cow-keepers throughout the eighteenth century, with some families farming there for decades. More herds were kept in Knightsbridge, Chelsea and the eastern fringe of Hyde Park.[14] The prevalence of cow-keeping in Southwark and Lambeth (14 per cent of operators identified) is partly explained by the proximity of pasture in St George's Fields and around Newington, but also by the area's industrial activities, discussed below. Geographers have observed that in the early phases of industrialisation, animal husbandry proliferates in suburban belts around consumption centres but that 'once living standards, environmental awareness and institutional capacity permit' these activities move away from the city.[15] Previous studies have assumed that this withdrawal was well underway in the eighteenth century but even in the late Georgian period, industrialisation and urbanisation did not force London's cow-keepers into the countryside. Rather, as the city expanded into the countryside, many operators benefited from their proximity to an expanding urban market. This complicates the orthodox assumption that with urbanisation, residential, commercial and industrial land uses supplant agricultural ones. It also throws into question whether or not metropolitan cow-keeping can be defined as a purely agricultural activity.

Cow-keepers adapted to urbanisation in remarkable ways, including reacting to dramatic increases in the cost of pasture. The rental value of suburban grassland increased from around £2–3 per annum in the 1780s to as much as £15 in the first half of the nineteenth century. Retreat into the countryside, where land rates were lower, was tempting but problematic as there was no effective means of transporting milk over distances of 8 km or more, without it spoiling or spilling from containers. While turnpiking improved roads in and around London in the eighteenth century, vehicles remained unsuitable for the carriage of milk and so the distance that a woman could walk with a heavy yoke and pails continued to determine where milk production took place.[16] Unable to flee the city, London's cow-keepers were forced to find urban solutions, a situation that transformed their trade. Some adapted by expanding their herds and engrossing land; in Islington and Clerkenwell, for instance, the number of operations peaked in the 1770s but then fell dramatically in the 1780s as the West, Rhodes and Laycock families began to monopolise.[17] Most cow-keepers, however, survived by scaling down, relocating and altering feeding regimes. Newspaper adverts for the lease and sale of cow-keeping sites suggest that the amount of pasture in use fell from an average of fifty-four acres in 1720–49 to forty-five acres in 1780–99. Likewise, sale of stock adverts indicate that the average size of herds fell from forty-five in 1750–79 to thirty-three in 1780–99.

The final quarter of the century also witnessed a significant geographical shift in cow-keeping activity. Herds ebbed away from Marylebone and the eastern edge of Hyde Park as fields were devoured by fashionable streets and land values soared. Meanwhile, cow-keepers multiplied in the poorer districts of Bethnal Green, St Pancras, Southwark, Rotherhithe and Deptford, a trend which began to undermine the trade's respectability.[18] These developments were entwined with a gradual transition from grazing to stall-feeding. In the 1750s, the vast majority of London's milch cows spent more than half of the year grazing but towards the end of the century this began to change. The growing need to replace grass with grain, hay and vegetables meant that for increasing periods cows were stall-fed in covered sheds constructed within or on the edge of the city. In the early 1790s, five London cow-keepers interviewed for the *Annals of Agriculture* confirmed that they kept their animals indoors for six or seven months of the year. At that time, many herds remained in fields overnight in the summer months but from the early nineteenth century they were increasingly returned to sheds at dusk. The move towards intensive stall-feeding was gradual and uneven but by the 1830s the traditional summer grazing period of early May to October was shrinking to six weeks in some cases and some animals never ventured beyond their walled enclosures.[19] After 1800, therefore, Londoners would have started to notice fewer cows grazing in nearby fields. This shift to yards and sheds curtailed some opportunities for Londoners to interact with cattle but, as we will see, it created others and the connection between milk production and consumption remained strong well into the early nineteenth century.

If historians have largely overlooked these developments, pig-keeping has fared even worse, with early modern descriptions of town-raised animals being useless, as well as agents of filth and nuisance, continuing to frame perceptions. Tobias Smollett's depiction of the London pig as 'an abominable carnivorous animal, fed with horse-flesh and distillers' grains' in *The Expedition of Humphry Clinker* (1771) is archetypal.[20] Smollett was primarily poking fun at rural prejudice, but pigs have come to epitomise the assumed incongruity of livestock-keeping in a civilised city.[21] Historians have also tended to assume that pig-keeping fell into decline in most parts of London in the seventeenth century and that throughout the Georgian period the city relied heavily on pork and bacon being carted in from the countryside.[22] This is not quite the case. The late 1600s did bring a new round of legislative action against porcine nuisance, both reinforcing existing orders and extending their reach. In 1671, an Act of Common Council restated a statute of 1562 to decree that 'no Man shall feed any Kine, Goats, Hogs, or any kind of Poultry, in the open Streets' of the City, but this did not prevent pigs being kept in enclosed yards.[23] Similarly, in Westminster, hogs were banned from wandering in the streets during the plague of 1582 but the keeping of swine itself continued to be permitted. A century later, pig-keepers deemed to be breaking nuisance laws were being prosecuted on a regular basis, suggesting that there were large numbers of small-scale operators at this time.[24] The authorities clamped down in the 1690s with London Street Acts which prohibited the keeping, feeding or breeding of pigs in any paved areas of the metropolis or within fifty yards of any building. The Middlesex justices printed 30,000 abstracts of the acts to be distributed in London's extra-mural parishes, which emphasises their determination but also the prevalence of pig-keeping in these districts.[25]

This makes it tempting to assume that pigs had been purged from London life by the early eighteenth century but in 1720, the Middlesex Justices were again forced to take action, this time deciding that once information against a pig-keeper had been given on oath, they would issue warrants to the churchwardens, overseers or constables of the parish to search for 'any Such Swine'.[26] This action was almost certainly provoked by fear surrounding the outbreak of plague around Marseille in 1719. In 1722, a survey presented to the Middlesex court identified straying hogs and putrid slaughterhouses among an array of serious nuisances affecting the city's streets. Yet neither this or subsequent outpourings of concern led to the eradication of pigs from the metropolis.[27] In 1762, Mayor Fludyer was forced to warn the 'many Persons' who continued to 'breed, feed, or keep Swine within this City and Liberties' that their animals would 'be seized and sold for the Use of the Poor'. Six years later, however, the *Public Advertiser* reported that 'a great Number of Swine' were still being seized in Holborn Upper Division'.[28] And as late as 1799, the vestry for St Clement Danes was forced to hold a special meeting 'to consider and give directions for removing the Hogs kept in Several streets in this Parish to the great Nuisance of the Inhabitants'. The churchwarden presented a list of known pig-keepers and the vestry clerk was ordered to write to each of them demanding they remove their animals within ten days or face prosecution.[29]

It becomes clear that throughout the Georgian period, the ambiguity and uneven enforcement of the law left a wealth of opportunity to keep pigs in many parts of the capital. Thus, when a pig-owning victualler from West Smithfield was asked at the Old Bailey in 1794: 'You know that pigs are not to be kept in [the City of] London?'; he responded, seemingly unfazed: 'Upon my word mine is a very large yard.'[30] Even when prosecutions were brought, the threat of confiscation and fines failed to break the bond between Londoners and their hogs. Whereas London's demand for beef and lamb relied on animals being driven in from the countryside, pork was predominantly an urban product. In 1822, 20,000 pigs were sold at Smithfield Market, but by then the city was annually consuming more than 210,000 hogs and 60,000 suckling pigs.[31] A small proportion of these were killed in the Home Counties and their carcasses carted into the city, but this was discouraged by the threat of deterioration and financial loss, particularly in summer. Many more pigs were fattened in the city than were driven in from the countryside because they were difficult to manage on the road and shed weight too quickly to remain profitable.[32]

Metropolitan pigsties are difficult to trace in this period, not least because they rarely appear in surviving architectural plans. Some of the most useful information can be gleaned from the Old Bailey Proceedings, where pig-keepers occasionally appear as victims of theft or the presence of these animals is referred to by witnesses in other trials. By searching the Proceedings online using the keywords 'pig', 'hog' and 'stye', I have been able to identify sixty-four active sites between 1730 and 1830. Clearly, this only represents a tiny fraction of the total and we cannot draw definitive conclusions about a constantly evolving practice with such a modest dataset. It is also important to note that Southwark's pig-keepers are excluded because the area's pig theft cases were heard at the Surrey Quarter Sessions (the area receives further attention below).[33] Nevertheless, it is both significant and unsurprising that the highest proportion of sites recorded in the Proceedings, nearly two-fifths, appear in the same northern suburban zone in which the city's milch cows were most prevalent, and that the East End (the Minories, Whitechapel, Mile End and Stepney) accounts for over a fifth of cases. The Proceedings also shed light on sties in Holborn, Old Street, Tothill Fields, St Giles, Soho and Hyde Park Corner, which emphasises that regulatory campaigns did not drive all animals out to the urban fringe. No less striking, however, is the apparent dearth of activity in the Square Mile, the only exception being a sty belonging to the Ram Inn which was located in West Smithfield, at the very edge of the City boundary and at the heart of the livestock trade.[34] While there is plenty of evidence that pigs were prevalent in medieval London, porcine nuisance cases are conspicuously absent in eighteenth-century records of the Mansion House and Guildhall Courts, despite both regularly hearing cases involving bullock 'hunting', mischievous dogs and equine traffic offences.[35] This supports the view that the City successfully regulated against pig-keeping but it is also likely that increasingly densely packed buildings and heavy traffic denied the space needed to erect sties and enable pigs to forage.

Drawing on other evidence, it becomes clear that pig-keeping in Georgian London developed in pockets of intensive activity surrounded by much larger areas which maintained a largely pig-free environment. Nuisance law helps to explain why. Opposition to pig-keeping occasionally led to court action, evidence of which appears in the records of the Court of King's Bench, which received the presentments of juries serving in Middlesex and, to a lesser extent, the City of London. These formal statements, made on oath, briefly describe the nature of crimes in cases where guilty verdicts were reached, along with the name, address and occupation of the guilty party. I examined a ten-year sample of presentments, covering the years 1735–37, 1759–63 and 1790–91, and found just six relevant cases.[36] While accusation rates would have been higher, this figure remains remarkably low. This does not indicate that there was a lack of pig-keeping, rather that operators usually avoided prosecution. The presentments suggest that olfactory nuisance was the predominant cause of complaint against pig-keepers. The case against John Jolly, a butcher of Holles Street, St Clement Danes, was typical: in 1760, Jolly was found guilty of keeping hogs near dwelling houses and feeding them with 'offals & entrails of beasts & other filth', all of which produced 'noisome smells'.[37] To prove a common nuisance, it had to be shown that a substantial number of people were being affected. Stench was the most likely nuisance to do so and represented an established tort, meaning that prosecutors prioritised olfactory evidence in court.[38] Thus, when Frederick Tasman, an Islington milkman, was presented in 1791 for permitting ten pigs to 'run up and down the Kings common highway', this was condemned not because the animals were obstructing traffic or rooting up roads but because they were exposing a large residential area to 'divers noisome and offensive smells'. Between the 1730s and the 1830s, however, metropolitan pig-keepers would have encountered growing leniency in court as industrialisation and urbanisation forced compromises in 'the standard of reasonableness' being applied to polluting manufacturers.[39] It is also important to remember that Londoners were habituated to the smells of livestock to an extent that many urban dwellers in developed countries would find difficult to imagine today. As Henry Mayhew noted, visitors often described the streets as smelling like a stable because of the vast number of horses and the huge quantities of dung which they produced.[40] London's air was pungently infused with a plethora of animal smells, competing for dominance with coal smoke and other man-made pollutants. Thus, many Londoners were prepared to accept the smell of pigsties providing that their noses and stomachs were not overwhelmed; and while it was widely assumed that bad smells could cause sickness, animal husbandry was not as closely associated with serious disease as it would be after the cholera outbreaks of the mid-nineteenth century.

The acceptability of a pigsty was highly dependent on its situation, scale and management, as shown by the case of Lewis Smart, a distiller and pig-keeper in the early 1730s. Smart's plot near Tottenham Court Road was, he claimed in court, ideal for a piggery because its environs were already blighted by the stench of cows, nightmen's pits, common laystalls and a ditch. Yet, as several witnesses testified, the stench from his sty carried to nearby Great Russell Street and its respectable residents complained that they were unable to sit in their front rooms because of the smell and that they had fallen sick. With good reason, Smart made the distinction between 'Erecting hogstyes in ye middle of ye town, and hogstyes in the outskirts of ye town' and complained that 'people build their house up to' existing pigsties. Smart failed to sway the jury and his conviction may appear to support the view that urbanisation forced agricultural activities out of the metropolis.[41] But a remarkable entry in the minutes of the Commissioners of Sewers and Pavements for Holborn and Finsbury emphasises that local circumstances were key.[42] In August 1773, the jurors of the sewer court set about investigating whether local pig-keepers were emptying dung into the Turnmill Brook and obstructing its passage. This activity was not new: there were sties in the area in the early 1600s and by 1683, the brook was already being 'choaked up' with dung. The sewer court jurors presented twenty offenders

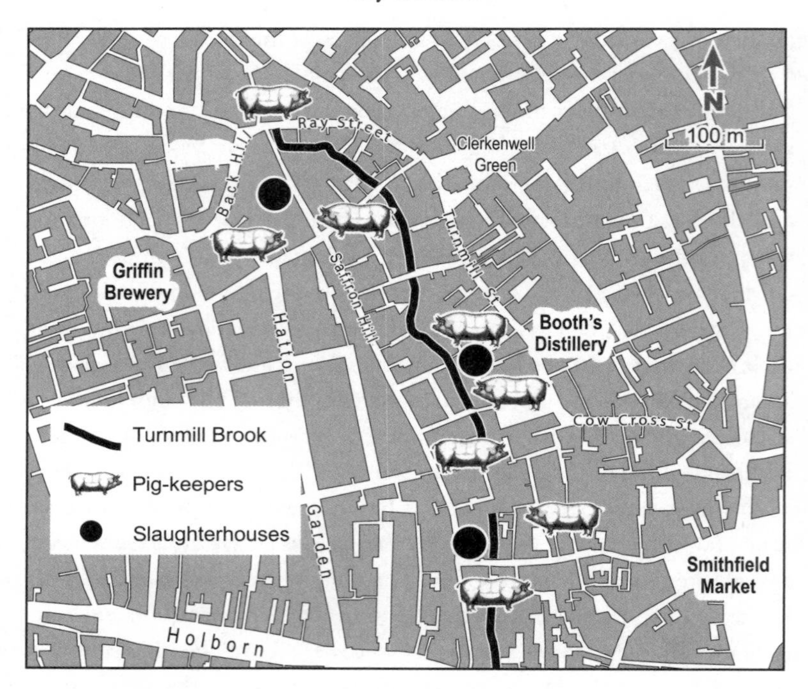

15 Locations of pig-keepers and slaughterhouses fined for emptying animal waste into Turnmill Brook in 1773.

operating at nine sites, each bordering the sewer (see Figure 15).[43] The jurors were only concerned with those who interfered with the sewer but powerful opposition to pig-keeping in nearby Holborn and the City hemmed operators into this narrow strip of land. Located on the approach to Smithfield livestock market, here was an island of agro-industrial activity involving four interrelated noxious trades: brewing, distilling, pig-keeping and slaughtering.[44] The sewer commissioners issued a substantial fine of £10 but it would have been difficult to prosecute these pig-keepers under nuisance law because their business was legitimised by its longstanding presence and the concept of aptness of place. In early modern and later Georgian London, the law tended to restrict pig-keepers and other polluting trades to pockets of land, where their activities exerted an intense influence on local life. Turnmill Brook is a compelling example but similar clusters can be found in Southwark, Tower Hamlets, Tothill Fields and elsewhere.

Having established that metropolitan pig- and cow-keeping were concentrated in a ring of built-up outer districts and urban fringe, it is important to note that these were the very same areas that saw some of the city's most impressive population growth. By the early 1700s, it is possible that half of the metropolitan population was already living in London's eastern and northern suburbs, as well as south of the river. This emphasises that livestock remained a familiar and influential feature of metropolitan life throughout the Georgian period.[45]

We have already seen that cow-keeping adapted to industrialisation and urbanisation in terms of the size and location of herds, but both the milk and pork trades underwent even more dramatic transformations in this period. In the twenty-first century, the industrialisation of meat production faces heavy criticism in relation to animal welfare, traceability, fraud, zoonotic diseases and food miles. Some of these issues are new but others have deep roots in Georgian discourse. The quality of town milk and pork received growing criticism in the second half of the eighteenth century. In the 1770s, for instance, Tobias Smollett caricatured London's milk as 'the produce of faded cabbage-leaves and sour draff' which milkmaids exposed to the city's filth. Such claims were not without foundation but the difference between urban and rural milk production was exaggerated.[46] The agricultural revolution intensified animal husbandry across the English countryside and for much of the period the treatment of urban milch cows bore considerable similarities to that of rural herds. Grass continued to dominate the diets of metropolitan herds into the early 1800s and in both town and country, farmers combined this with a mixture of foodstuffs which included hay, turnips and swedes, as well as vegetable refuse. In the capital, much of this was grown locally by suburban market gardeners who, in turn, acquired urban manure. As the metropolitan milk trade expanded, cow-keepers began to purchase vegetables from a wider hinterland and by the 1830s, some large operators even maintained their own supply farms 16 km or more from the city.[47] Throughout the Georgian period, cow-keepers across Britain experimented with feeding regimes designed to boost milk and meat yields, but in London a distinctly urban solution was found in the city's colossal drink industries.

After extracting liquid wort from grains, brewers and distillers were prepared to sell their waste product, 'spent grain', to local cow-keepers. The transformation of an industrial by-product into a low-cost, protein-rich animal feed gave these urban farmers the key to unlocking higher milk yields and greater profits. Crucially, the brewing season lasted from October to May, meaning that an abundance of wasted grain arrived precisely when the city's cows were most in need of stimulating food.[48] In the 1750s, it was observed that on Sunday evenings, cow-keepers' carts, 'for three Miles round this Metropolis' could be seen driving 'through the Streets, to fetch the Grains from the respective Brew-Houses they deal with'.[49] Some operators made direct contracts with breweries – in the early 1800s, William Clement was purchasing grain from Charrington's Brewery in Mile End and hired carters to make regular deliveries to his yard on the Hackney Road – while others used grain merchants as middlemen, some of whom were fellow cow-keepers.[50] In each case, as William Youatt observed, 'The dairyman… must know his brewer, and be able to depend on him' to ensure the supply was fresh, unadulterated and reliable.[51] Sometimes, these agro-industrial relationships produced shared sites; in 1782, for example, the *Gazetteer* advertised the sale of a plot in Bethnal Green, where the 'farmer and cow-keeper' Pearce Dunn kept milch cows on premises which he shared with a dealer in yeast and stale beer. In common with many other operators, Dunn also maintained a pigsty, into which he recycled the protein-rich whey from his herd's milk.[52]

As access to pasture declined and milk production became more intensive, London's cow-keepers became increasingly reliant on industrial partnerships. By the early nineteenth century, wasted grains accounted for 20–35 per cent of an average cow-keeper's expenditure on feed and leading concerns achieved impressive economies of scale by bulk ordering.[53] In 1810, the Laycock family employed eighty horses to draw fifty carts; and in 1819, Mr Millan, the owner of a 'large Milk Farm' in Paddington, ordered around 23 cubic metres of grain to be carted across the metropolis from Whitbread's brewery every day.[54] By contrast, the majority of smaller cow-keepers were forced to cluster around large breweries and distilleries to minimise haulage costs; this process was well underway by the 1740s in at least three major cow-keeping districts. John Rocque's plan of the city shows that cow-keepers in Hoxton and Shoreditch were served by at least two distilleries and two breweries less than 1 km away. The herds of Tothill Fields abutted five breweries, including the enormous Stag brewery on Castle Lane; and in Southwark, there was at least one distiller and four breweries.[55] The parish of St Saviour, Southwark, provides particularly striking evidence of the intensification of this agro-industrial relationship in the late eighteenth and early nineteenth centuries. Figure 16 plots the locations of fourteen cow-keepers, as surveyed for the vestry in 1807.[56] By then, Southwark's cowsheds had moved into the heart of its industrial zone,

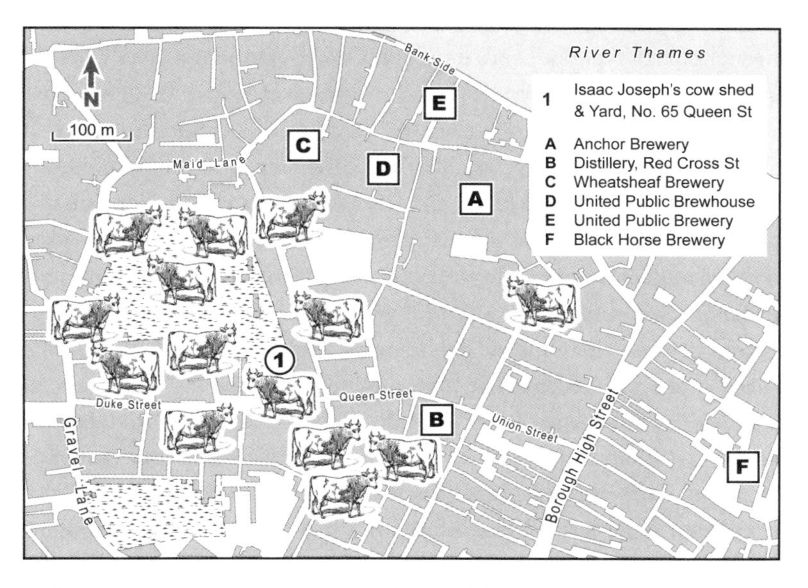

16 Locations of cow-keepers in St Saviour parish, Southwark, in 1807.

occupying yards within a few hundred metres of London's premier brewery, the Anchor, as well as two other substantial breweries on Maid Lane, the Wheatsheaf and the United Public. Together with a distillery on Red Cross Street, these sites provided an inexhaustible supply of food for Southwark's herds.[57] Meanwhile, distilling also underpinned dramatic changes in metropolitan pig-keeping.

A formidable omnivorous digestive system meant that pigs could be fattened on a panoply of cheap foodstuffs readily available in Georgian London and this versatility gave them a symbiotic niche in the metropolitan economy. While historians have noted that farmers experimented with pig feeding in the English countryside, particularly in the second half of the eighteenth century, it is often forgotten that Londoners had been doing so for centuries.[58] A 1697 husbandry manual which includes 'Instructions to fatten Swine in Towns' admitted that urban dwellers lacked 'the advantage of leting [sic]' pigs 'run abroad so much' but celebrated their access to cheap, flesh-raising foodstuffs. These ranged from vegetables which 'may be bought for little matter' to offal, whey and the 'Washings of Ale Barrels'.[59] Pigs continued to devour this swill throughout the period and developed a reputation for being useful animals because they recycled large quantities of urban waste – matter which scavengers would otherwise have had to collect and dispose of at a significant cost – into profitable flesh. This helped to feed the metropolitan population, particularly the city's less wealthy majority, but also satisfied much of the Navy's demand for salted pork. Pig fat was also used in the manufacture of candles and soap, for which there was fast-growing demand. Even as streets and buildings swallowed up green space, pigs thrived because their spatial requirements were relatively low, enabling small-scale operators to profit from modest backyards. At the other end of the scale, however, metropolitan distillers fattened huge numbers of pigs in vast sties on yeast deposits, or lees, and other waste from the production of gin.[60] This was already happening in the 1600s but the practice accelerated in the early eighteenth century. In 1736, a defender of distillers and therefore someone keen to downplay the scale of their activities, claimed that the number of hogs fattened 'does not exceed 50,000'; but in 1783, it was estimated that the total had been closer to 100,000 during the 'gin craze' of the 1720s–50s.[61] Whatever the true figure, by the late 1730s, the scale of these activities was such that London distillers were forcing farmers in Shropshire and the Home Counties out of the market. Unable to fatten their animals as cheaply, some of these rural pig-keepers signed petitions in 1740 and the House of Commons considered the findings of a Committee investigation in 1745. There would be little respite, however, until the 1760s when increasingly harsh duties on distilling reduced pig stocks significantly, perhaps to as few as 30,000.[62]

Pig-keeping may have been steeped in urban tradition but conditions in Georgian London created new forms operating on an unprecedented scale. This is underlined by developments in the second half of the eighteenth century when a handful of large distillers came to dominate the trade. In 1748, the Swedish agricultural economist Pehr Kalm recorded that London's distillers often held between 200 and 600 animals and sold them to butchers 'at a great profit'.[63] But when Thomas Pennant visited Lambeth's 'vast distilleries' in the 1780s, he found 'seldom less than two thousand hogs constantly grunting'. A few years earlier these sites had been the property of Sir Joseph Mawbey, a leading metropolitan distiller and MP for Southwark and Surrey, who was mercilessly mocked by James Gilray for his porcine activities.[64] Even more impressive operations included Thomas Cooke's distillery at Milbank, Johnson's at Vauxhall and Benwell's at Battersea, which each fattened as many as 4,000 pigs. In the 1760s, Cooke's animals were consuming 350 tonnes of grain every month and demanded the attention of five men. While smaller distillers sold their hogs alive to carcass butchers, Cooke invested in a sophisticated processing plant to slaughter, cut and cure his fattened animals on-site, enabling him to profit from the lucrative trade in finished bacon. In March 1767 alone, the company slaughtered and cured 324 hogs. The remarkable scale on which some London distillers kept pigs in this period was not just unprecedented, it remained unsurpassed in British farming until the twentieth century.[65]

As well as gaining privileged access to London's voracious market for pork and pig fat, distillers benefited from their proximity to the Navy's victualling yards at Tower Hill and Deptford, where they secured valuable contracts based on discounted bulk orders.[66] This helps to explain the high concentration of pig-keepers in the Minories, as well as in Whitechapel, Bethnal Green and Mile End. In 1776, the Whitechapel distiller Samuel Liptrap contracted with the Navy Board for £8,200 to supply 2,000 hogs, which he delivered in six batches of around 300 animals over nineteen days. Liptrap probably fattened some of these animals himself but almost certainly sourced animals from other sties.[67] This lucrative deal illustrates the central role which London's distillers played in the supply of pig flesh to the Navy, then Britain's biggest industrial client.[68] Archaeological excavations have shown that the operations of the Tower Hill victualling yard ballooned in the eighteenth century: in the late 1720s, a new slaughterhouse the size of two tennis courts was constructed, as well as a hanging house with capacity for around 700 carcasses, new hog pens, a scalding house and a cutting house. The yard remained in use for sixty years and succeeded despite being far removed from the major Channel ports, in large part because of its proximity to a copious, reliable and cost-effective supply of pigs.[69]

Thus far, this chapter has argued that pig- and cow-keeping were dynamic sectors of the metropolitan economy but their influence went far beyond this. Keith Thomas assumed that only a narrow group of English men – butchers, colliers, farmers, grooms and cab-drivers among them – were 'directly involved in working with animals' in the eighteenth century and that even fewer owned the animals themselves. Thus, he argued, 'the vast majority of urbanites' were 'remote from the agricultural process and inclined to think of animals as pets rather than working livestock'.[70] But this overlooks a wealth of evidence for London which shows that a broad spectrum of lower middling and plebeian men and women were engaged in animal husbandry. Moreover, changes in the geography and organisation of these activities promoted diverse interactions between people and livestock. It is to these entanglements of human and animal lives that we now turn.

Domestic cow- and pig-keeping are generally associated with the countryside, in part because they played a vital role in the English cottage economy in the early modern period but also because they became politicised in the nineteenth century. In the 1820s, William Cobbett highlighted the absence of a cow or pig outside a labourer's house as evidence of rural hardship and the proletarianisation of farm workers. Cobbett passionately defended the cottager's right to raise livestock while casting London, the 'all-devouring WEN', and its idle 'tax-eaters', as parasites on the countryside and presenting its industrial workers as the antithesis of the self-sufficient cottager.[71] Londoners certainly siphoned off the fat of the land in the Georgian period but Cobbett obscured the fact that thousands of men and women in the city did keep cows, pigs and poultry. Metropolitan cow-keepers came from a wide range of occupations, including some as unrelated as shoemaking. In most cases, however, the two occupations were symbiotic, with cow-keeping often taking the lead. Several cow-keepers in the northern suburbs, for instance, took advantage of the clay deposits beneath their fields to manufacture bricks and tiles.[72] For others, however, cow-keeping was the sideline. Dealers, chapmen and butchers were well placed to keep cows because they knew how to negotiate the livestock and provender markets. The same applied to carmen and coachmasters, who also had access to yards and the vehicles needed to transport food and dung. Meanwhile, victuallers benefited from close relationships with breweries and distilleries to secure favourable terms on spent grain.[73] But the business was open to any tradesmen able to invest a few pounds in a cow and access a modest area of pasture. This is illustrated by the involvement of rope-makers. The trade's use of strips of land known as walks, mostly located in fields around Shadwell, Rotherhithe and Limehouse, gave master rope-makers a valuable opportunity to depasture cattle and even to erect cowsheds. Insurance records show that in the 1780s, William Cornwell's property in Sun Tavern Fields, Upper Shadwell, included a timber building which served as a hemphouse and a cowhouse.[74] Notwithstanding these signs of entrepreneurial success, it has to be acknowledged that lower middling and plebeian Londoners seeking to enter the cow-keeping business faced mounting obstacles.

At the beginning of the eighteenth century, there were numerous common pastures on the outskirts of the city, including those in St George's Fields, Stoke Newington, South Lambeth, Woolwich, Wandsworth, Clapham and Chelsea. Encroachments on these lands by enclosure and urban expansion began in the seventeenth century but gained momentum in the Georgian period. These developments suggest that access to grazing rights became more restricted but also that there was significant variation across the metropolis.[75] Until the late eighteenth century, St George's Fields, land owned by the City of London, retained a strip system of agriculture and was open for common grazing between Lammastide and Candlemas, although the extent to which these rights were exercised, and by which social groups, is difficult to establish.[76] Decades before parliamentary enclosure acts swept grazing rights away, metropolitan commons rarely offered the urban poor an opportunity to keep their own cow. Nevertheless, some commons did play a significant role in metropolitan husbandry for much of the period. In Woolwich, ancient grazing rights were exercised throughout the eighteenth century on an eighty-acre common owned by the Crown. Around 1760, the Woolwich vestry successfully defended this activity by blocking building leases and when the Board of Ordnance began using the land for artillery practice in the 1770s, local residents refused to remove their cows. It was only in 1803 when the Board asserted its authority by statute that Woolwich's rights to herbage were finally extinguished.[77]

By comparison, domestic pig-keeping was far more accessible. One reason for this was that pigs did not require pasture. They also reproduced rapidly: sows were fertile after just one year, compared to more than three years for a cow, and could deliver a litter of several piglets after just four months, whereas a cow usually only produced one calf after nine months.[78] Just as importantly, pigs fattened cheaply and efficiently: in the 1780s, a piglet worth 3 or 4s could be transformed into a well-fattened hog worth £2 within a year.[79] Drawing on probate inventories, Carole Shammas has suggested that rates of pig-keeping in the East End of London doubled from 1.6 to 3.5 per cent between 1661–64 and 1720–29, bucking the decline seen in smaller towns in the south of England. This would be impressive but, as Shammas acknowledges, the data for the earlier period is limited and that for the later period is skewed towards wealthier individuals.[80] It may not be possible to pin down precise numbers but the prevalence of live pigs, as well as chickens and rabbits, among items stolen from plebeian and lower middling households suggests that cottage-style husbandry was common in Georgian London, particularly in outer districts.

Victims' testimonies in the Old Bailey Proceedings provide a window onto this way of life. They reveal, for instance, that in 1774, William and Ann Collard were living with a pig, hen and five young chickens in their house in Bunhill Row, in the built-up parish of St Luke's, Old Street. Fifty years later, we learn that a family in Shoreditch kept eight fowls and fifteen rabbits in their washhouse, and that Mr Ireland of Tottenham Court Road was feeding several pigs on 'things used in the house' and 'emptied into the trough'.[81] It is significant that in three-quarters of the pig theft trials identified, the crime took place in the immediate vicinity of the owner's residence. While large-scale piggeries may be underrepresented because they were more secure, it is fair to assume that the vast majority of pig-keeping Londoners were small-scale operators owning, the Proceedings suggest, fewer than five animals.[82] At all scales, pig-keeping remained a subsidiary activity. Rarely in eighteenth-century England did men give 'pig-keeper' as their occupation, at least not for the purposes of insurance or probate.

Pig-keeping was an attractive and accessible sideline for a wide variety of working Londoners but particularly the city's artisans, shopkeepers and tradesmen. Those identified in the Proceedings include a carpenter, stone mason and baker, as well as victuallers and chandlers. As one writer observed in the 1830s, pigs were 'especially valuable to those persons whose other occupations furnish a plentiful supply of food at a trifling expense'.[83] It is unsurprising, therefore, to find that William Barrow, a scavenger in 1770s Westminster, kept three pigs in a sty close to his house to recycle the kitchen scraps, market waste and other edible matter which he scraped from the city's streets and yards.[84] The processing and marketing of food were two of the biggest occupational sectors in the metropolitan economy; the city teemed with inns, taverns, chop-houses, pie shops and bakeries, and each of these businesses provided a potential niche for pigs to serve as living recycling plants. Thus, in the 1790s, we find that the keepers of a chandler's shop in East Smithfield and a 'cook-shop' in Brick Lane were each fattening four pigs. This was not new – London's bakers had been keeping pigs for centuries – but the numbers of operations and pigs involved were unprecedented.[85] Meanwhile, many of the city's victuallers took advantage of substantial yards to erect sties and treated their animals to lashings of stale beer and food scraps. We know about this because these well-fed pigs attracted so much unwanted attention: in 1785, John Thompson was robbed of two large hogs valued at £4 from his inn, the Pied Bull at Islington; and in 1794, William Goodall of the Ram Inn, West Smithfield, apprehended a man stealing a pregnant sow which he had bred himself.[86]

Evidence from the Old Bailey Proceedings emphasises that the majority of pig-keepers lived in close proximity to their animals. House and sty were often close enough for owners to hear their animals 'grunt' or 'squeak' as thieves disturbed them. In outer districts such as Clerkenwell, pigs were often let out of their sties during the day and allowed to roam in their owner's backyard. Some animals were even permitted to enter their owner's living quarters: in 1797, an elderly brick-maker's servant living in Battle Bridge told the court that a 'porker', which he had bred, 'used to run about the house'. And in some instances, urbanisation appears to have forced people to permanently share their homes with pigs. In 1814, a resident of Hoxton admitted to keeping an adult pig in a sty erected 'next [to] the kitchen' because he 'had no yard'; and in 1829 a man in Deptford claimed to have kept a sow 'at the back of my house' for 'about a year and a half'.[87] These cases may not have been typical but even when house and sty were separated, most pig-keepers came into close contact with their animals at least twice a day. Several victims of hog theft described feeding their animals in the evening when they secured the sty and checking on them again in the morning, at which point some animals were let into the yard to roam. These testimonies provide fleeting glimpses of entwined lives and help us to understand complex attitudes to livestock in this period.

While cattle, sheep and pigs probably did become more commodified in Georgian Britain, the pace of change can be exaggerated and evidence of emotion and sentiment overlooked in everyday interactions. First and foremost, these animals were viewed as investments but relations based on money can also be infused with emotion, imagination and sentiment. Sows could be kept for up to six years for breeding and their young fed up for several months, meaning that small-scale pig-keepers got to know the physical and behavioural characteristics of their animals remarkably well.[88] It is hardly surprising, therefore, that pig-keepers found it so easy to recognise their animals when they were stolen, even when they had already been killed. In 1796, Francis Newland of Mile End Old Town recalled that his pig had 'a few hairs of a different colour at the end of the tail... A rusty colour' and affirmed 'I bred her myself'. When challenged 'Have you not seen 10,000 pigs' like this, he replied, 'I never saw a pig like my own; I know her by her size and make'. Similarly, in 1811, a cow-keeper in Bagnigge Wells was able to identify two carcasses in Worship Street, telling the Old Bailey, 'I am very certain they were my pigs. One was a black and white sow, and after they had been scalded the black marks still remained.'[89] One testimony crashes through the boundary between business and pleasure: in 1803, John Willis, a gentleman 'in a confidential situation in the Bank' and a livery stable-keeper living on the Hackney Road, explained 'I have for some time past, for amusement when at home, kept a few rabbits, and some pigs'.[90] Willis may have been eccentric but it was not uncommon for pig-keepers to express what we might tentatively describe as fondness for their animals.

Domestic pig-keepers did not regard their animals as pets but when they were kept in backyards, entire families became closely involved in their care, and the work of feeding, watering and mucking out involved close observation, communication and physical contact. The pigs would have greeted the arrival of their owners with grunts and squeals; and given the opportunity, they would have nudged shoes, legs and hands with their snouts; and stood still to have their backs scratched. Underlying these interactions was a powerful bond of dependency: just as pigs relied on their owners for food and security, families had an invested interest in the wellbeing and security of their animals. We can glean a sense of this from the testimony of William Hatfield of Park Street, Kennington. In 1827, he recalled 'I saw my sow safe… while I was cleaning the sty… I turned her into the yard; she cost me 5l. before she pigged: when I had done the sty I could not find her' and, in common with several other victims, he stressed, 'she never strayed'. Hatfield had invested in his sow, cared for her and allowed her to roam in his yard. In return, she produced piglets and repaid his trust by never straying. Porcine sagacity clearly influenced how owners viewed and treated their animals; in court, some were emphatic that their pigs could find their way home, even from a thief's den in another parish. Others pointed out that their pigs recognised and followed them, behaviour that reinforced a sense of mutual trust and reliance.[91]

The bond between Londoners and their pigs remained strong until the mid-nineteenth century, when sanitary reformers launched a sustained attack on working class districts, where the practice appears to have intensified significantly in the early 1800s. In Kensington's Potteries, for instance, pigs may have outnumbered people by as many as three to one by the 1840s.[92] Around the same time, the physician and sanitarian Hector Gavin visited the East End and condemned the stinking pigsties which 'abound everywhere', while Friedrich Engels was appalled to find that in Manchester, the Irishman 'eats and sleeps' with his pig and lets 'his children play with it'. For Engels, this was a despicable example of Irish behaviour infecting England's industrialising towns, but as we have seen, co-habitation with pigs flourished in London long before the mass migrations of the nineteenth century.[93] Indeed, migrants from the countryside had been bringing agricultural skills and rural ways to London for centuries. For these newcomers, pig-keeping provided a sense of cultural continuity as well as an opportunity to make money. But more than this, pig-keeping was part of a lively interchange between urban and rural economic activities which saw workers moving freely between the city and its rural hinterland. Many farm labourers in the London basin, for instance, spent time working in the capital's manufacturing trades or as middlemen in the food trades, before returning to the fields when demand for labour returned.[94] Pig-keeping continued to enjoy a high level of acceptance in Georgian London, retaining strong links to respectable urban trades and lower middling domestic economies. While pigs were condemned for causing nuisance, they were also useful and many Londoners successfully defended their activities on these grounds.

We have seen that men and women from diverse occupations engaged in pig- and cow-keeping but these activities also ensured that virtually the entire metropolitan population interacted with livestock in one way or another. Urbanisation often brought people and livestock closer together in this period, rather than pushing them apart. As more cowsheds were erected in the vicinity of houses and commercial properties, a melting pot of middling sorts, tradesmen and labourers came to live or work cheek-by-jowl with the city's herds. Striking evidence of this appears in the 1807 survey of lands and buildings in St Saviour, discussed above. By the turn of the nineteenth century, Southwark's manufacturing zone was surrounded by residential and commercial properties, and as shown in Figure 16, it was among these houses and shops that the majority of cow-keepers operated. They were not the only operations involved in the livestock trade – there were several butchers' shops in the area, each with their own slaughter houses – but this was not an area dedicated to raising and slaughtering animals. An array of properties and occupational groups abutted the cow yards. At No. 65 Queen Street, for instance, Isaac Joseph's cows were surrounded by residential properties; nine shops including those of two chandlers, a basket-maker, a barber, an apothecary and a cheesemonger; as well as three alehouses. Everyone who lived, worked and shopped in Queen Street would have seen, smelled and occasionally touched these animals.[95]

Newspaper and Old Bailey trial reports also show that Londoners entered cow sheds and yards for many different reasons: some adults, for example, went looking for lost property or climbed drunkenly into haylofts to sleep, while children turned cow yards into hazardous playgrounds.[96] It becomes clear that many Londoners were highly confident around livestock in the eighteenth century, not least the city's thieves. Despite being far less portable than pigs or poultry, and more dangerous to handle, cows remained attractive. In the summer months, thieves painstakingly stole milk from grazing animals, by hand.[97] And in more serious cases, cows were driven out of fields and sold to butchers; by 1789, the crime had become so common in Camberwell that the vestry was forced to offer a £10 reward for the conviction of those 'stealing any horse, bullock, cow or sheep from any field or common'.[98] Even more frequently, animals were driven to local pounds in exchange for an illicit reward. In the 1730s, the City's Court of Aldermen, ordered that 'persons who bring in stray cattle shall give their true name and address to the Keeper of the Green-Yard pound' in Cripplegate, 'and shall not be rewarded until 48 hours later'.[99] Nevertheless, the practice continued into the nineteenth century as did bullock hunting, a raucous plebeian sport that involved goading Smithfield droves until enraged animals charged through crowded streets (see Chapter 4).

More respectable forms of interaction with cows stemmed, in part, from a distrust of milk-sellers. Some discerning customers demanded to see their milk being taken from the cow to ensure its authenticity and this gave rise to new modes of retailing in the second half of the century. Some cows and asses were walked to the front doors of wealthy residences to be milked but more often cow-keepers invited customers, or their servants, to visit their premises.[100] Towards the end of the eighteenth century, dairy shops were constructed in front of yards to improve this experience. George Scharf depicted such an arrangement in 1825 as part of a study of London's evolving shopscape. His drawing of a cow-keeper's shop in Golden Lane, on the northern edge of the City, not only records its old-fashioned facade but also the presence of several cows waiting to be milked behind a low partition adjacent to the counter (see Figure 17). While much more refined than a cow yard, this layout still enabled customers to see, hear, smell and even touch the animals supplying their milk. As suggested in Chapter 1, the connection between animals and semi-luxurious goods was both accepted and often actively promoted in this period. While trade cards celebrated the contribution of mill horses in the production of goods including paint and snuff; barbers, apothecaries and perfumers also invited customers and their servants to watch 'bear grease', a pomade for hair and wigs, being cut from recently slaughtered brown bears to authenticate this expensive product.[101]

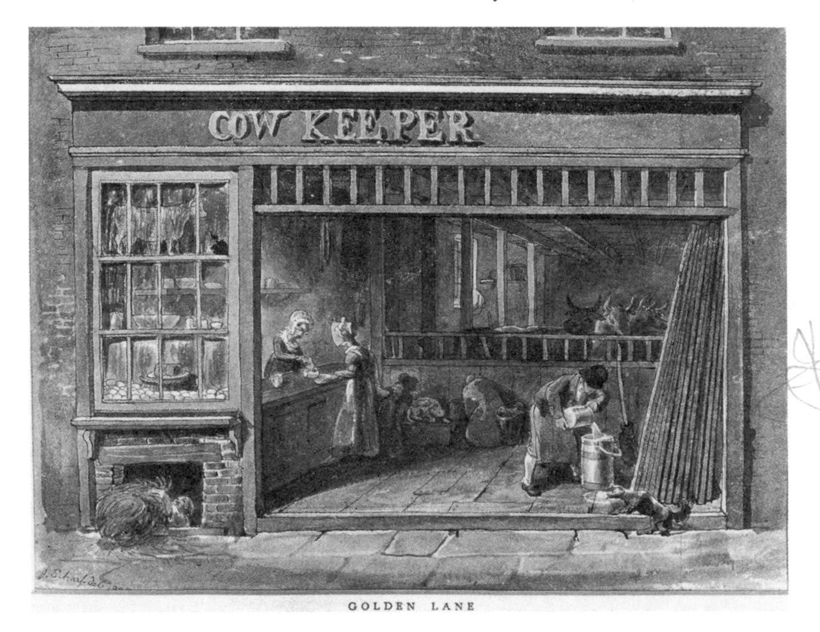

17 George Scharf, *A Cowkeeper's shop in Golden Lane* (drawing, 1825).

Consumers placed at least some pressure on cow-keepers to keep their animals healthy. If not out of sympathy for the animals, then for their own wellbeing, customers did not want to drink milk from filthy or diseased beasts. We should not, therefore, assume that criticism of cow-keeping in the 1850s reflected attitudes or actual conditions in the Georgian period. While some cowsheds were almost certainly overcrowded in the eighteenth century, intensification accelerated after 1800 as cows began to live most of their lives indoors. Scharf wrote that he selected the Golden Lane site to compare it with the later Westminster Dairy, 'an elegant Milk Shop in the Quadrant, Piccadilly'.[102] In doing so, the artist drew attention to changes in shop decoration in the Regency period but even more striking is the disappearance of milch cows from public view. This marked the start of an accelerating separation of milk consumption from production in the Victorian period.

Quality control was not the only motivation for respectable Georgian Londoners to get close to milch cows. These animals appealed to polite society's taste for *rus in urbe* and played a significant role in recreational life. Milk became fashionable in this period partly because of growing artistic and literary interest in the pastoral and a belief that milk 'offered a tonic' for those weary of urban life.[103] At the same time, it was a response to living in an agropolis, a celebration of London's bounty. The capital's parks, pleasure gardens and suburbs were popular venues in this period, offering relief from the hubbub of the city, and interaction with milch cows was an integral part of the experience. Topographical views of the city and its environs often depicted polite Londoners walking through fields and meadows among herds of languorous cattle. In the foreground of Thomas Bowles' engraved prospect of London from Islington, published in 1730, a respectably dressed man can be seen entering a fenced enclosure to fish in the Upper Pond and appears perfectly at ease with the cows resting nearby.[104] Similar but increasingly idealised images appeared in the late eighteenth century, by which time the spread of brickfields and dust-heaps, combined with the gradual withdrawal of grazing herds, had begun to erode the pastoral qualities of London's urban fringe.[105] In 1772, the satirical poem *Town Eclogues* complained:

> Where'er around I cast my wand'ring eyes,
> Long burning rows of fetid bricks arise,
> And nauseous dunghills swell in mould'ring heaps,
> Whilst the fat sow beneath their covert sleeps.
> I spy no verdant glade, no gushing rill,
> No fountain bubbling from the rocky hill,
> But stagnant pools adorn our dusty plains,
> Where half-starv'd cows wash down their meal of grains.
> No traces here of sweet simplicity,
> No lowing herd winds gently o'er the lea,
> No tuneful nymph, with cheerful roundelay,
> Attends, to milk her kine, at close of day.[106]

These lines accurately associated urbanisation with the intensification of animal husbandry but throughout the Georgian period, Londoners continued to enjoy convenient access to unspoilt countryside where they could walk, ride, hunt, fish and pursue other rural diversions. At Strawberry Hill, 16 km from the city, Horace Walpole wrote in 1752 that he had enjoyed 'a syllabub under the cow', by which he meant milking a cow into a glass of cider or ale. Walpole considered this one of the benefits of leading 'quite a rural life' but milk could be purchased fresh from the udder in several of London's parks and pleasure gardens.[107] The most fashionable *lactarium* was in St James's Park, where the resident herd began to inspire poets and artists in the first half of the eighteenth century.[108] Visiting in 1765, Pierre-Jean Grosley was delighted to find milk 'served, with all the cleanliness peculiar to the English, in little mugs at the rate of a penny a mug'. This experience is brought to life by a vibrant conversation piece painted in the mid-eighteenth century (see Figure 18). In the immediate foreground of *St James's Park and the Mall*, a cow rests in the shade of a tree as several well-dressed men, women and children gather around a milkmaid's stall. This image usually receives

18 Detail of *St James's Park and the Mall*, attributed to Joseph Nickolls (oil on canvas, after 1745).

attention because it depicts polite sociability but the interactions between people and cows are usually overlooked. Grosley, however, appreciated that these animals were key to the park's appeal, commenting that they gave its 'walks a lively air, which banishes solitude from them when there is but little company'.[109] Consuming milk in this way was in sharp contrast to the artificial shopping experience offered by the city's fashionable shops. The appeal of park milk depended on its associations with the natural and the pastoral but the authenticity of this experience remains questionable. The St James's Park herd may have occupied an idyllic glade but it was one managed by gardeners, divided by footpaths and surrounded by bustling streets.

The pleasure gardens of suburban Islington and Marylebone were perhaps more convincingly pastoral. On a pleasant August evening in 1767, Sylas Neville, a respectable bachelor lodging in Bloomsbury, unwound at Islington's White Conduit House. At half-past eight, he 'drank a pint of milk warm from the cow' which, he later wrote, 'I would do oftener, had I not so far to go over the stones to get at it'. Neville then looked out over the fields towards the city and observed 'What thick air we breathe in London! The smoke appears like an immense cloud over it, when seen from the country.'[110] This is an intriguing statement but it is worth comparing Neville's perceptions to an earlier depiction of Islington by William Hogarth. *The Four Times of the Day: Evening*, painted in 1736, depicts a cow being milked on a patch of land dividing Sadler's Wells theatre and the Sir Hugh Middleton, a busy tavern and popular place of resort (see Figure 19). Some historians have argued that the cow in this image emphasises Islington's 'near-rural setting'.[111] But bearing in mind the area's fast-growing population, as well as its prominent role in the milk, livestock and leisure trades, I would argue that what this image really shows is that urbanisation brought Londoners and livestock closer together. Hogarth depicts a family of affluent citizens taking an evening stroll within a few metres of a milch cow and takes comic advantage of this to place the horns of a cuckold above the patriarch's head. The cow's swollen udders mirror the condition of the man's heavily pregnant wife and Hogarth may even be likening the woman's unborn bastard to the

19 Bernard Baron after William Hogarth, *The Four Times of Day, Plate III: Evening* (engraving, 1738).

milch cow's calf, a creature which farmers often viewed as an inconvenience. Seen in this light, Islington stages a characteristically Hogarthian study of urban life, morality and the entangled lives of people and animals.

This chapter set out to show that Georgian London was a thriving agropolis and to challenge the notion that Londoners became agriculturally unproductive in this period. We have seen that cow- and pig-keeping were significant sectors of the metropolitan economy which adapted to and profited from urbanisation. We have also seen that interaction with livestock was far from limited to farmers, drovers and butchers because a broad spectrum of plebeian and middling families kept livestock and an even more diverse group encountered cows and pigs simply because of where they lived, worked and played. As productive as London's farmers were, however, they were far from able to satisfy the city's enormous demand for animal flesh. The expansion of London's hinterland was crucial to feeding the city's growing population and this brought an immense trade in meat on the hoof to the streets.

4

Meat on the hoof

> Fat droves of sheep, consign'd from *Lincoln* fens,
> That swearing drovers beat to *Smithfield* penns,
> Give faint ideas of *Arcadian* plains,
> With bleating lambkins, and with piping swains…
> But droves of oxen through yon clouds appear,
> With noisy dogs and butchers in their rear,
> To give poetic fancy small relief,
> And tempt the hungry bard, with thoughts of beef.
>
> Charles Jenner, *Town Eclogues* (1772)[1]

For the poet Charles Jenner, the Smithfield livestock trade epitomised London's metamorphosis from the bucolic muse of his predecessors into the frenetic metropolis of his own day. Jenner accurately describes his city greedily consuming animals from the furthest reaches of the country and its streets swarming with cattle and sheep, but this was not a new phenomenon. The trade in meat on the hoof had evolved over a period of more than 800 years and by the 1740s, well over half a million sheep and 80,000 cattle were being sold in the city every year. Despite this, historians have shown far greater interest in the demise of London's inner-city livestock market in 1855 than in its 900 years of trading.[2] This is a missed opportunity because the story of London's trade in meat on the hoof is crucial to understanding the city's development, its relationship with the rest of the British Isles, its ecological footprint and the lives of its people. In 1856, one Londoner reflected that the trade had been 'a continued struggle against difficulties, almost against impossibilities; a continued protest against the dictates of good sense' as well as 'a continued display of the meat-buying powers of the London public; and… a sort of perennial declaration of the wonderful improvements gradually introduced in the size, quality, and condition of grazing stock'.[3] This chapter examines each of these claims and argues that the Smithfield livestock trade was one of the most impactful economic activities in the Georgian metropolis.

Consumption, as well as production, played a critical role in Britain's Industrial Revolution and metropolitan consumer behaviour was uniquely influential. London provided the principal shop window for an ever-expanding range of commodities but historians generally focus on innovative manufactured goods and exotic imports.[4] Georgian London's voracious appetite for agricultural produce has received attention but almost entirely from a rural perspective. It has been shown, for instance, that the capital was instrumental in the development of more market-orientated agriculture, as well as enclosure and rural depopulation, changes in provincial market culture, and popular protest in rural areas.[5] These are convincing and valuable arguments but the seismic forces exerted by the livestock trade on metropolitan life demand equal consideration. The systems by which livestock are marketed and killed are central to the relationship between those animals and their consumers. For this reason, William Cronon has described the opening of Chicago's New Unified Stockyard in 1865 as a watershed in the populace's detachment from nature. 'In a world of ranches, packing plants, and refrigerator cars', Cronon argues, most of 'the constant reminders of the relationships that sustained one's own life… vanished from easy view'. Once Chicago received its meat from hundreds of kilometres away by rail, it seemed 'less a product of first nature and more a product of human artifice'.[6] As we will see, the situation in Georgian London was dramatically different. Here, the sale, distribution and slaughtering of livestock remained an intra-urban process and as the trade expanded, the population's exposure to cattle, sheep and pigs became increasingly intense.

Meat on the hoof

Scholarly interest in financial institutions and West End shops appears to have overshadowed the fact that Smithfield Market was one of the most successful centres of commerce in Georgian London.[7] Beef, lamb and to a lesser extent pork were devoured in huge quantities by the elite, but also by London's middling sorts and lower orders, albeit in smaller amounts per capita. Fascination with London's meat consumption inspired numerous attempts to calculate its scale and progress. In 1694, for instance, John Houghton estimated that Londoners consumed 100,000 cattle annually, based on the assumption that each family 'shall in a year eat one beef' but admitted that this was conjecture. Two months later, he published a revised estimate of 88,400 based on an 'ingenious' butcher's assertion that around 1,700 cattle entered London each week.[8] Even this probably exaggerated by a few thousand animals and anticipated levels first reached in the 1720s or 1730s. Later studies attempted to calculate the weight of meat consumed in the city – in 1815, one suggested a grand total of almost 159,000 metric tons a year and in 1856, per capita consumption was said to have increased from 32 kg to 55 kg between 1750 and 1850 – but these figures should also be treated with caution.[9] We can be much more confident about the numbers of livestock being sold in Smithfield because these were recorded by the clerk of the market. In 1725, therefore, we know that 76,531 cattle changed hands in the market square and by 1786, this had risen 41 per cent to 108,075.[10]

Meat epitomised the unique intensity of metropolitan consumption. When it came to demand for farm output, all English cities were important but as contemporaries observed, London was unique in that its demands far outweighed the size of its population. In 1784, the agricultural writer Arthur Young explained:

> At first sight, it may seem that the same people dispersed would carry with them their markets and their demands: but this is not the case. It is the union to a spot, the concentration of wealth that is alone powerful to give that impulsive motion that is felt at the very extremities. Such a city as Bristol can form communications by road or navigation to a distance of a few miles: It is a vast capital only that can extend them to the extremity of a kingdom.[11]

As the focal point of the nation's political, financial, legal and cultural life, the metropolis housed a much higher proportion of affluent consumers than any other city. These individuals not only consumed greater quantities of meat than people in other parts of the country, they also paid higher prices for better cuts from 'more highly finished' animals.[12] Moreover, as well as serving Londoners, Smithfield supplied huge quantities of meat to the Navy. In October 1746 alone, the Victualling Board ordered contractors to supply its main processing centre at Deptford with 1,600 cattle within a month; and in 1828, an experienced Smithfield salesman recalled that in wartime 'a great buyer' had driven '5 or 600 Lincolnshire Cattle... through the City' every week for the Navy.[13]

By the mid-eighteenth century, Smithfield Market occupied four and a half acres, expanding to six and a quarter acres in 1834. Its agricultural activities filled much of the week but the sale of bullocks, sheep, lambs, calves and hogs took place on Mondays and Fridays. Sheep, of which there were generally seven times more than cattle, were by far the most numerous while lambs, calves and hogs appeared in much smaller numbers: in 1822, for instance, 149,885 cattle, 1,507,096 sheep, 20,020 pigs and 24,609 calves were sold.[14] Between the 1730s and the early 1790s, the average quantity of sheep brought to market rose by a quarter while cattle increased by a fifth.[15] This is all the more remarkable considering that rinderpest is thought to have destroyed over half a million animals in Britain between 1745 and 1768.[16] In 1803, the sixth Duke of Bedford informed Parliament that average annual sales of sheep and cattle in Smithfield had increased by 200,000 and 30,000 respectively since the early 1790s. This would suggest that the trade's rate of expansion doubled in the late eighteenth and early nineteenth centuries. These figures should be taken with a pinch of salt as they served the duke's campaign to enlarge Smithfield Market but the expansion of London's livestock trade in the Georgian period was undoubtedly impressive and far outstripped developments in any other European city.[17] By 1790, London's consumption of livestock was at least three times greater than that of Paris, Europe's second largest city, which consumed between 150,000 and 200,000 animals.[18] By 1809, the number of sheep and lambs annually slaughtered in London had already surpassed one million and in 1822, Smithfield was processing an astonishing 1.7 million animals per year (1,507,096 sheep, 149,885 cattle, 24,609 calves and 20,020 pigs), all transported on the hoof through the city.[19] Cattle sales at Smithfield only reached their peak of 277,000 in 1853, just two years before the trade was moved to suburban Islington.[20] To put this in context, on its final day of trading, 11 June 1855, London's inner-city trade in meat on the hoof dwarfed that of contemporary Chicago, the giant new city which already commanded thirteen railroads. One visitor to Chicago's inner-city stockyards celebrated their 'astounding dimensions' and in 1861, 177,000 head of cattle were driven through the city's streets.[21] London handled 100,000 more.

Meat on the hoof was big business in Georgian London, dwarfing many other boom markets which have received far greater attention from historians. In the 1780s, the value of meat sold in the capital equalled, if not exceeded, that of Britain's total imports (£3,613,147) from its sugar colonies. The relatively new trades in sugar, rum, tea, coffee, cocoa, cotton, chinaware and exotic hardwoods were no match for London's ancient trade in cattle, sheep and pigs.[22] By 1809, the Smithfield livestock trade was thought to be worth £5 million pounds, by which time a single market day could generate £100,000 of sales. By 1815, its annual value had risen to £6.68 million and by 1832 it stood at £8.5 million.[23] The Georgian metropolis developed a highly sophisticated livestock economy which both benefited from and encouraged advanced commercial practices. In 1813, it was observed that the 'Landed and Grazing Interest expect their remittances to be forwarded by Post the same Day the Stock is sold'. Such swift transfers were made possible by a 'system of credits and promissory notes' which connected 'farmers, graziers, drovers and salesmen in the country to the City'. These developments were closely linked to the emergence of country banks after 1750 which used an agent or a correspondent bank in the capital to process bills of exchange. By lubricating the wheels of the economy with short-term credit, this system underpinned the success of medium- and long-distance trade.[24] Smithfield's proximity to London's financial district and the horse-drawn postal carriers operating out of Bishopsgate, Aldersgate and Holborn was critical. By the early nineteenth century, Smithfield's 600 salesmen enjoyed convenient access to seven or eight banking houses in the immediate vicinity of the market and in 1828, a 'Money-taker and banker at Smithfield' warned against moving the market even two miles out of London because, he said, 'we have great intercourse all the day with the Bank and the City bankers, and... the hour of the payment is the hour of the market'.[25] This helps to explain why, for almost a century, repeated demands to move the trade out of Smithfield were so ineffective. Another significant factor was the pressure exerted by local businessmen who relied on the trade's continuation, not least the several innkeepers who kept 'the large accommodation necessary for the people who attend the market'. The Corporation of London also profited from the trade through tolls and market dues, but this remained a relatively modest revenue stream and does not adequately explain its defence of such a disruptive activity.[26] For all its sophistication and success, the trade posed serious challenges for the city and impacted heavily on daily life. The trade's scale, geography and organisation maximised interactions in the streets of the metropolis and continually clashed with the ideals of enlightened urbanity.

Located just outside the City walls, Smithfield had been in use as a suburban cattle market since 950 AD. In 1300, it was still set in open countryside but by 1700 it lay at the heart of a heavily populated commercial hub. The trade was experienced far beyond the confines of the marketplace. London's expansion in the eighteenth century not only increased demand for meat, it meant that livestock had to be driven over greater distances through ever more congested streets. In the mid-eighteenth century, approximately 1,500 cattle and 11,500 sheep were brought to Smithfield each week, shared across the two market days. By the 1820s, this had risen to 3,000 and 28,500 respectively.[27] Livestock droves were, of course, prominent features of many Western European and North American cities in the eighteenth and early nineteenth centuries but counterintuitively they were far more deeply rooted in London life than they were in smaller cities. Numerous attempts were made to move livestock markets and slaughterhouses out of English cities in the early modern period. These were often met with strong resistance but nowhere more so than in London. In 1488, Henry VII banned the killing of beasts within the walls of London but persistent opposition from the city's butchers led to the act being repealed in 1532.[28] For the next 300 years, the Smithfield trade would continue to resist interference and in 1798, an exasperated humanitarian lamented that

> Even were the whole Court of Aldermen, to be tossed by horned cattle, their united influence would not be able to carry such a measure, as the removal of Smithfield Market. A man might as well have the modesty to ask for universal suffrage, and the abolition of the slave trade.[29]

This immovability was in complete contrast to the fate of livestock markets in the rapidly expanding industrial cities of Manchester and Liverpool. Between 1780 and 1837, Manchester's cattle market was forced from the heart of the city to the outskirts and then 5 km further afield to the satellite town of Salford. Liverpool's first substantial cattle market was established on the outskirts of the city in the 1780s but was soon relocated to Kirkdale, a township 5 km to the north of the city, before settling in the village of Old Swan, 5 km to the east, in 1830. By then, a 'moderate' market in each of these provincial cities involved 5,000–6,000 animals, whereas London processed more than 30,000 per week.[30]

Beyond the British Isles, it is worth comparing London to Paris, a far smaller city, which had virtually no tradition of urban livestock marketing in the early modern period. In the early thirteenth century, the law directed that the city could only be supplied from the markets at Sceaux, 10 km away, and Poissy, 24 km away. In 1416, Charles VI proceeded to prohibit the driving of livestock through the streets of Paris and constructed four slaughterhouses on the outskirts of the city. This system continued throughout the eighteenth century and in 1809, Napoleon constructed five more suburban abattoirs to keep the driving and slaughtering of animals out of Paris.[31] New York presents another stark contrast to London. In 1830, North America's most populous city remained seven times smaller than the British metropolis but Manhattan's livestock markets and slaughterhouses had already been shut down, leading one English visitor to enthuse: 'This city is now supplied from the State of New Jersey, Long Island, and the rivers above and below New York.'[32] The fact that London retained its gigantic inner-city livestock market decades after smaller cities had removed their own is indicative of the trade's unique economic and cultural potency.

The geography, population size, economics, governance and culture of eighteenth- and early nineteenth-century cities fostered remarkably different conditions for the sale and distribution of meat. What is most striking about London is the extraordinary degree to which the city forced its populace to interact with livestock and how deeply the business of meat on the hoof was rooted in the wider metropolitan economy. The night preceding market day, cattle and sheep were collected from suburban pens which encircled the metropolis. From outposts at Islington in the north; Holloway and Mile End in the east; Knightsbridge and Paddington in the west and Newington in the south, droves began to converge on the heart of the city until they became a dense swarm.[33] In the first half of the century, most drovers approaching from the West went along Oxford Street before making their way through Holborn. This ensured that man and beast encountered some of the city's heaviest pedestrian and horse-drawn traffic. Once completed in 1756, the Paddington to Islington New Road sought to alleviate the worsening congestion and nuisance generated by livestock droves and horse-drawn waggons.[34] Residents of the parishes of Saint George Hanover Square, St James Westminster, Saint Ann Soho, Paddington and Saint Marylebone petitioned in favour of the proposed bypass, insisting that it would 'prevent the frequent Accidents and Obstructions that happen by Conveying [cattle] two miles or upwards through the paved streets'.[35] Support also came from those wanting to safeguard the interests of Holborn's waggon services: the obstructions caused by the great number of animals being 'constantly drove through Holborn' were, they claimed, 'a great hindrance' to 'Trade and the Dispatch so essential thereto' while cattle 'running wild about the streets' caused frequent accidents.[36] Concerted calls to banish cattle droves from the West End had emerged in the mid-1750s as part of a much broader street improvement campaign, but support for the New Road shows that the interests of polite improvers were not necessarily opposed to those of commerce.[37] In April 1756, nearly 200 graziers, salesmen, butchers, drovers and dealers in cattle who attended Smithfield signed a petition in favour of the plan, asserting that it would enable them

> in a more Expeditious manner and with much greater ease and security bring their Cattle… quiet and cool to and from the Market and transact the other necessary Business of their Employments which are of such General concern to the Publick.[38]

Thus, the New Road not only promised to bring greater order and safety to the streets, but also to expedite trade and commerce. In practice, however, it appears that at least some drovers refused to use the New Road, presumably because it added nearly 1 km to their journey.[39]

Drovers from the north entered the metropolis via Highgate and Angel, before heading into the City along St John's Street. By the early 1700s, both areas had become important staging posts – in 1754 alone, 267,565 sheep and 28,692 cattle passed through the Islington turnpike – and developed distinctive bovine cultures as a result.[40] Some inns were given symbolic names such as the Pied Bull or proudly displayed bullock horns. Some of Islington's leading milch cow-keepers, discussed in Chapter 3, offered food, water, shelter and security for cattle and sheep before market days. By the 1820s, Richard Laycock could accommodate 2,000 beasts, which generated a substantial annual income of around £7,000.[41] South of the Thames, drovers arriving from Kent, Sussex and Surrey amassed on the Old Kent Road. Prior to the Lambeth Enclosure Act of 1806, a patchwork of commons and wasteland provided a final resting place for livestock, before they were driven across London Bridge and steered through the City's narrow streets. As these droves converged on the narrow approach roads to Smithfield, the area became a seething tide of life. Finsbury Square was close to the eye of the storm and in 1828 a jaded resident complained that 'from eleven till four o'clock in the morning there is one uninterrupted scene of noise and confusion' and that by midnight, the area was 'in an uproar with Drovers'.[42]

On arrival at Smithfield, sheep were guided into pens concentrated in the centre and north-west corner of the market. Around half of the cattle were fastened to rails opposite St Bartholomew's Hospital while the other half stood closely packed together in 'off-droves' near Smithfield Barrs. Once sold, drovers led some animals directly to Smithfield's slaughterhouse district, a hive of compact facilities located behind or in the basements of shops run by independent retail butchers.[43] Many more beasts were forced to walk further to their deaths. The first stopping point was Hatton Garden, a wide street where cattle were divided up for the final leg of their long journeys. Brook's and Fleet Markets were only a stone's throw from Smithfield but six other sites were 1.5–3 km further west. The closest of these were Clare and Bloomsbury Markets, followed by Soho's Newport and Carnaby Markets. Beasts intended for the Grosvenor, St James's and Shepherd's Markets, in the south-west corner of the city, were driven furthest and experienced the full extent of London's variegated topography and street life.[44] As we will see, this procedure tested the nerves of both livestock and drovers but also guaranteed that virtually all Londoners came into regular contact with livestock.

Most working Londoners rose shortly after six o'clock in the morning and began work just before seven o'clock, precisely when livestock were being driven out of Smithfield. To compound matters, Monday, the busiest market day, had been a day off work for many Londoners in the first half of the eighteenth century but thereafter this practice fell into rapid decline, which meant that many more people encountered livestock as they went about their business.[45] The West End elite generally rose too late to encounter this morning traffic but they could not avoid the trade entirely as droves were still at large in the streets well into the early hours of the afternoon. Shopping excursions were frequently disrupted by errant cattle: in August 1765, for instance, the *Lloyd's Evening Post* reported that a bullock had run into the shop of Mr Jackson, a salesman in Holborn, and was 'removed from thence with great difficulty, after breaking the glass of the back parlour to pieces, and doing other considerable damage.' By the early nineteenth century, such incidents were so common that they had inspired, and brought into common parlance, the phrase 'like a bull in a china shop'.[46] More commonly, however, it was terrified pedestrians rather than cattle that barged into shops from the street. In 1828, an upholsterer and cabinet-maker on Ludgate Hill complained that every other week 'we have not less than five or six ladies coming in the shop to avoid the cattle'.[47]

Consumer behaviour appears to have magnified the impact which livestock droves exerted in the city. Throughout the period, Saturday was the busiest day of the week for the sale of meat, primarily because many families bought beef or lamb for their Sunday lunch.[48] The inconvenience of purchasing cattle on Friday to kill immediately for Saturday's sale meant that butchers preferred to buy on Monday and keep their purchases alive until Thursday or Friday morning. This allowed sufficient time to cut the meat while also retaining its freshness. This practice was commercially astute but resulted in Monday markets becoming dangerously swollen and the city's streets being overwhelmed with unpredictable animals.[49] Once sold, some livestock were temporarily incarcerated in sheds or slaughterhouse basements – in the 1820s, for instance, William Collingwood, a butcher in Newgate Market, kept up to 400 sheep in a shed in nearby Red Lion Court – but some wholesale butchers also hired grazing grounds in the suburbs. As early as 1680, Richard Hodgkins, a former master of the Butchers' Company, held leases on pasture in Barking, West Ham, Plaistow and Woolwich. As discussed in Chapter 3, the rising cost of suburban land in the eighteenth century may have forced more butchers to resort to cramming their animals into sheds and basements but there were commercial incentives to avoid this. Allowing livestock to graze for several days or, if possible, weeks enabled them to regain the bulk which they had lost on their exhausting journeys and thereby fetch a higher price. For this reason, Collingwood still sent some of his sheep to graze in a small field near Battle Bridge while his peer, Valentine Rutter, used a shed in Goswell Street before transferring his flock to the nearby Artillery Ground, a patch of grass flanked by houses, in Finsbury.[50] These practices meant that cattle and sheep not only lived in London for longer but also spent more time on its streets, being driven between suburban holding pens, grazing grounds, butchers' sheds, markets and slaughterhouses.

Modern research into the behaviour of domesticated livestock helps to explain why conditions in Georgian London would have been challenging for these animals and their guardians, the town drovers. Sheep display a strong tendency towards allelomimetic behaviour, which lends itself to herding because they instinctively follow members of the flock initiating movement and walk in a column. But these natural instincts came under enormous strain in the metropolis. Sheep are easily startled by the sudden appearance of shadows, reflections or unexpected sounds, and will alert other members of their flock to potential danger by raising their heads and adopting a tense stride. While strong flock affiliation benefited drovers when their sheep were together, once an animal became separated it would often panic and become difficult to manage. Moreover, if an isolated sheep caught sight of its flock, its instinct was to run towards it regardless of any threat, including a horse-drawn vehicle, in its path.[51] The transfer of animals from country drovers to town drovers almost certainly made managing them in the city more difficult because sheep can recognise individual humans and would have been more reluctant to follow an unfamiliar new herdsman. A view of *Soho Square*, published in 1812 (see Figure 20), demonstrates that the task was challenging even on a quiet street. The drover which it depicts has managed to keep a dozen sheep together in a loose column but the leaders have begun to deviate, perhaps because they have been spooked by the two bullocks in the same drove. Erratic ovine behaviour was a continual source of frustration for drovers and helps to explain why some individuals beat errant animals, as depicted by William Hogarth in the *Second Stage of Cruelty* (see Figure 2).

Cattle were even more difficult to manage because they wielded far greater bulk and power. For this reason, in 1828, an experienced Smithfield salesman told a parliamentary committee investigating the state of the market that moving cattle through crowded streets

20 Anon., *Soho Square*, aquatint published in Rudolph Ackermann's *The Repository of Arts, Literature, Commerce, Manufactures, Fashions and Politics*, vol. 8 (1812).

made 'a certain violence necessary to be used'. One of the committee members, the MP Edward Protheroe, sympathised, asserting that during a visit to the market he had seen 'instances of severity' but none that he 'considered very censurable, for the drovers were placed in the most harassing situations, in which had I myself been placed I should have acted with greater severity than they did'. A witness called to the committee, a land agent from Waterloo Road, agreed with the statement that 'those cruelties inflicted upon the cattle… a man with ordinary feelings is almost necessitated by the crowded state of the market to commit'. Similarly, a shoe warehouse keeper asserted that drovers used 'as little cruelty as they can' but admitted that when men are 'compelled to exercise this cruelty, they will not be very delicate of the manner in which they use it, after a length of time'.[52] The committee did not hear from any drovers but we can be confident that all would have considered striking cattle necessary to secure their co-operation in the metropolis. At the same time, however, these men understood that irritating or frightening a bullock could provoke a devastating response.

Despite their size, power and, in many cases, deadly horns, the bullocks being driven through London were often described as being timid by nature. The Islington cow-keeper, Richard Laycock, advised the 1828 committee that 'the animal is more frightened at the public than the public at the animal'.[53] When accompanied by other animals in a compact group, bullocks could be commanded fairly safely but as soon as they felt alone or exposed, they became extremely anxious.[54] Criticising the livestock trade in the 1760s, the street improver John Gwynn observed that accidents

> are chiefly owing to the separating of these animals from each other, to which they have a natural aversion; when one of them is parted from the herd it always endeavours to recover his situation, but being prevented and finding himself alone, which he is unaccustomed to be, he runs wildly about… and at length from the natural principle of self-defence often does irreparable mischief.[55]

For this reason, two drovers charged with manslaughter in 1786 were heavily criticised for separating a bullock from a drove as they transferred the animal to a nearby yard.[56] Agricultural guides from the period observed that certain breeds displayed particular behavioural characteristics. William Youatt praised Galloway cattle for being 'very docile', noting that 'It is rare to find even a bull furious or troublesome… a most valuable point about them', whereas the Ayrshire bull 'was too furious and impatient of control to be safe'. Youatt judged that hornless cattle, 'being destitute of the natural weapon of offence', were 'less quarrelsome and more docile' but concluded that 'the ferocity of the horned beast is oftener the effect of mismanagement than of natural disposition'.[57] Some drovers held years of experience but the trade was constantly attracting young men who had to learn by trial and error. Between field, market and slaughterhouse, two or three drovers might take charge of a drove, a system which relied on the exchange and due consideration of information about animal behaviour. In the 1786 trial, the drovers were accused of proceeding into the open street 'well knowing the said bullock was wild and mischievous'. It was suggested that one of the defendants had been told that the bullock had run wildly at a drover in Lincolnshire but later claimed that it had behaved 'pretty well' in London and 'came very well along with the rest of the beasts'. The witness who gave this version of events, another drover, added that he had handled 'a bullock yesterday... as mad as a March hare almost' and swore 'we always tell a butcher when he is wild'.[58] Despite all such efforts, however, drovers knew from painful personal experience that no amount of knowledge or skill could guard against erratic bovine behaviour. When animals charged or swung their horns, they were often the first to suffer the consequences. In 1757, the *London Evening Post* lamented that as one of the drovers was untying an ox 'fastened to a Rail in Smithfield Market, it gave a sudden toss with its head, and jabb'd its horn into the drover's eye, by which means the poor man's eye dropped out of his head'. Ten years later, 'an ancient Drover' was gored 'so terribly' after endeavouring to free his animal from Red Lion Court that he died before reaching hospital; and in December 1789, the *London Chronicle* reported that a butcher had become the third member of his family in one generation to be killed by the same breed of bullock.[59]

As the population, trade and traffic of the metropolis increased, livestock were exposed to even more unnerving stimuli and denied the conditions upon which their co-operation depended. In their 1756 petition for the New Road, the drovers warned that even with 'the utmost care' they could no longer prevent their animals 'running wild, terrifying and often killing... the passengers in the streets' and complained that they 'sustained many losses', with animals being lamed or killed because of the great rise in vehicles, particularly in the districts bordering Smithfield.[60] Holborn was particularly problematic on account of serving the greatest concentration of waggons in the metropolis as well as accommodating major retail markets, the Inns of Court and hundreds of bustling shops, commercial yards and stables. But in the 1786 trial, one witness recalled that the bullock had been spooked by 'the carts and coaches [which] made such a noise' in Clerkenwell; and in the 1820s, a livestock salesman complained that in Smithfield itself, he had 'seen beasts with their claws taken off by the drays… and I have seen the hips of the oxen injured in the same way… I have had the coach poles run against me three or four times a day; I have enough to do to kept myself from getting under the waggon wheels'.[61] Conditions were equally challenging in the City's dense network of narrow streets and alleys. In the 1760s alone, the press reported four encounters with bullocks in the Royal Exchange, the first of which, in May 1761, led to some people 'losing hats and wigs, and some their shoes, while others lay upon the ground in heaps, with their limbs bruised'.[62] Such incidents highlighted the extreme incompatibility of two of the City's most profitable economic activities. In 1828, a resident noted that Monday was 'a great day of business in the City' during which 'there is a greater influx of individuals' and complained 'it is precisely on that morning that… the City is almost impassable from the cattle'.[63]

The challenge facing drovers was exacerbated by mischievous interference by passers-by. Legislative action, court proceedings and newspaper reports reveal that many Londoners made sport with cattle droves on Smithfield Market days. In his analysis of the City's summary courts, Drew Gray found that a remarkable 10 per cent of the 582 prosecutions for street-related regulatory offences recorded between 1784 and 1796 were for the abuse or chasing of cattle. Considering that offenders usually avoided prosecution, this data suggests that bovine trouble-making was persistent and widespread.[64] Drovers often complained that passers-by deliberately startled their animals with loud noises or sudden movements, prodded them with sticks or pelted them with stones. One of the drovers put on trial in 1786 pleaded in his defence that three bakers had 'rattled their pails' at the animal, 'making game of him' while a witness recalled seeing a man 'daring the bullock' by 'wavering his hat backwards and forwards'.[65] Spontaneous provocations of this kind were dangerous enough but cattle were also the subject of organised urban sport in this period. Bullock hunting was already popular in and around the City by the time Hogarth depicted 'a rabble of boys, and dirty fellows' chasing an animal until it 'maddens with rage' in Holborn in 1751.[66] The sport involved separating a bullock from a drove and, as a reformed participant recalled in later life, 'menacing the drovers and frightening the bullocks' by shouting, whistling and beating the animals with sticks. The timing and location of the attack were carefully chosen to cause maximum chaos, 'generally where two streets crossed' and often resulted in 'hunters', drovers or innocent passers-by being injured or killed. Hunts could go on for an hour or more, until the beast was too exhausted to run any further and could then be recovered by the drover.[67] In 1781, Parliament legislated that anyone who 'shall pelt with Stones, Brickbats, or by any other Means drive or hunt away, or shall set any… Dogs at any… Cattle, without the consent of the owner' should be arrested and fined.[68] The new legislation appeared less than a year after the Gordon Riots of June 1780, at a time when the authorities were seeking to clamp down on disorderly behaviour of all kinds.[69] In this climate, cattle were recast as dangerous weapons wielded by the most unruly elements of society. While the 1781 Act probably deterred some bullock hunters, the speed and confusion of the chase made it much more difficult to police these activities than bull baits, which the authorities had successfully forced into the suburbs by the 1770s, helped by the fact that they were generally advertised in advance and took place on a fixed spot.[70] The relative leniency of the punishments handed out to bullock hunters by the Guildhall and Mansion House justice rooms also helps to explain why the sport survived for so long. Of the fifty-seven prosecutions identified by Gray, around half led to fines, a third of offenders were discharged, and only three were imprisoned. While steadily diminishing, reports of bullock hunts continued to appear in London's newspapers into the 1820s and the sport may have survived in a more restricted form until the Smithfield livestock trade ended in 1855.[71]

While many Londoners acknowledged the difficulties which drovers faced, these men bore the brunt of growing anger and frustration. The vast majority of accident reports published in the London press described the cattle involved as 'over-drove', immediately implying that drovers, through 'carelessness or bad conduct', were solely to blame when their animals ran wild in the streets.[72] This assumption rested on an Enlightenment belief that men could both guard against accidents and be masters of nature.[73] Virginia DeJohn Anderson has shown that early modern English colonists of North America blamed domestic animals going wild on human interference and viewed livestock less as independent actors than as 'passive objects of human manipulation'. Thus, the wrongdoing of animals 'advertised their owners' failure to maintain control' or, worse still, their desire to create disorder.[74] This outlook was central to perceptions of the human–animal unit in Georgian London and shaped the policing of metropolitan streets in significant ways. From the 1760s, metropolitan newspapers sardonically labelled drovers as the 'Smithfield gentry' to suggest that these low-born men abused their command over powerful animals to challenge the city's social hierarchy. At the same time, they were described as 'two-legged brutes' and 'brutes in human shape', which implied that working with animals made them dim-witted and violent.[75]

As condemnation intensified, the authorities ratcheted up their regulation and punishment of drovers. These developments were closely linked to metropolitan improvement debates but also a growing desire to hold men more accountable for actions, however unintentional, which caused injury and death.[76] Many complaints centred on the perceived inexperience, incompetence and laziness of drovers which, it was claimed, increased the risk of animals becoming disorderly and magnified their ruinous potential. Drovers were accused of ignorantly and complacently relying on brute force to manoeuvre their animals, behaviour which tended to enrage cattle. A correspondent for the *Gazetteer* in 1764 proposed the complete prohibition of 'sticks, whips, and other weapons' to foster less cavalier droving techniques, while others criticised the use of bull-terriers to intimidate cattle. In March 1765, a widow gentlewoman was fatally tossed by a bullock, in pursuit of which, the *London Chronicle* complained, were 'three butchers dogs'.[77] Such incidents were grist to the caricaturist's mill: in *The Overdrove Ox* (1787), Thomas Rowlandson traced a trail of destruction on London Bridge to a bullock being pursued by a pack of dogs and stick-wielding drovers.[78] And in *Miseries of Human Life*, *c.* 1800, George Woodward depicted a gentleman hopping with terror between the horns of two bullocks which are being harried by a bull-terrier

21 George Woodward, *Miseries of Human Life: As you are quietly walking along in the vicinity of Smithfield on Market day finding yourself suddenly obliged though your dancing days have been long over, to lead outsides, cross over, foot it, and a variety of other steps and figures, with mad bulls for your partners* (hand-coloured etching, *c*. 1800).

(see Figure 21). Such impressions were not without foundation – excessive and misplaced use of sticks and dogs almost certainly provoked bovine disorder in some instances – but the harshest criticism of drovers typically came from those with the least experience of managing livestock. In 1774, Parliament passed an *Act to Prevent the Mischiefs that arise from driving Cattle within the Cities of London and Westminster*, thereby giving constables the power to arrest drovers suspected of 'negligence, or ill usage'. Convicted drovers faced fines of 5–20s, a serious penalty for men earning around 5s a day. The act also entitled the City to set down its own regulations for the driving of livestock and to impose fines, but the tragedies continued.[79] In 1781, Parliament felt compelled to strengthen this legislation by clamping down on bullock hunting and eleven years later, the City's Court of Aldermen published a new set of regulations

22 William Henry Pyne, *Smithfield Drover*, hand-coloured aquatint, published in Pyne's *Costume of Great Britain* (1804).

designed, according to the *Public Advertiser*, 'to keep the drovers of cattle in order'. The court banned the use of sticks 'below the hock' or with pointed goads more than a quarter of an inch long and ordered every drover 'to wear a numbered badge' on their arm or be fined 20s.[80] The latter measure appears to have had some success – in the early 1790s, the London press reported on the conviction of a handful of drovers 'for refusing to wear a badge' and in 1804, William Henry Pyne depicted a law-abiding drover wearing a badge marked with the City's arms and '127 SM' [Smithfield Market] in his *Costume of Great Britain* (see Figure 22) – but into the 1830s humanitarian campaigners complained that some drovers still worked without badges or wore forgeries.[81]

The Smithfield trade caused horrific accidents, injuries and deaths throughout the Georgian period. In the 1760s alone, newspapers reported on eighteen deaths from cattle and a further twenty-six cases in which the victim was said to have been 'carried off for dead' or similar.[82]

Enraged cattle tossed, gored and trampled Londoners in the streets, leaving them with broken ribs and limbs, fractured skulls, severe bruising and deep puncture wounds. In January 1758, the *Whitehall Evening Post* lamented that a bullock 'ran at a poor Man in Tooley Street, and drove him with so much violence against a wall, that his brains came out'. The following year, it reported that an ox had 'tossed several persons in Oxford Road and did considerable mischief, particularly to a well-dressed woman, whose left eye was gored out; and she received several other wounds. She was carried to the Middlesex hospital but without hopes of recovery'.[83] Historians have good reason, therefore, for describing the livestock trade as one of the worst nuisances and sources of barbarity in the Georgian metropolis but we should not lose sight of the fact that it was also one of its greatest achievements. No other city had ever marketed, distributed and slaughtered so many livestock across such a large and heavily populated area.[84] Smithfield's increasing notoriety did not mean that its operations were becoming antiquated or less profitable. And the trade continued to attract praise: after visiting Smithfield in the 1790s, a successful butcher from Bath declared that 'the business between the salesmen and butchers, was done in the most fair and honest manner'. He was 'astonished' to see the livestock put in their places so early in the morning and

> so nicely judged by the salesmen, who are not only the readiest men I ever noticed in dealing – selling to a wonderful nicety, amidst such crowds of cattle – but act according to the trust reposed in them, by far distant depending customers, graziers, or jobbers, &c.[85]

Moreover, the suggestion that modernity arrived with the ascendancy of dead meat transported by rail in the second half of the nineteenth century overlooks a long history of expansion and innovation in the supply of live meat to the metropolis.[86] As Miles Ogborn has argued, there are 'many ways in which modernity's spaces are produced' and Georgian London displayed a 'variegated topography of modernity' which included 'ambiguous' spaces of which Smithfield, I would argue, was one of the most influential.[87] While the trade exemplified nuisance and cruelty on the one hand, Smithfield Market and the livestock sold there were also totemic symbols of agricultural improvement and metropolitan prosperity.

Meat inspired patriotic pride across Georgian Britain but was celebrated with particular gusto in London. The city's unrivalled beef proved a succulent muse to metropolitan writers and artists, making its presence felt, for instance, in 'The Roast Beef of Old England', a ballad first performed in Henry Fielding's *The Grub-Street Opera* in 1731; and Hogarth's painting, *O the Roast Beef of Old England ('The Gate of Calais')*, completed in 1748. Meat also fuelled many Johnsonian witticisms: at the start of a 'good supper' at the Mitre tavern in July 1763, Samuel Johnson remarked that the rain outside 'is good for vegetables, and for the animals who eat those vegetables, and for the animals who eat those animals'. And in 1778, the great man, who 'boasted of the niceness of his palate', claimed that he could write a superior cookery book which 'would tell what is the best butcher's meat, the best beef, the best pieces'.[88] In the eighteenth century, beef and mutton were increasingly portrayed as foods of the urban artisan and tradesmen, as well as of the yeoman farmer. In reality, these meats remained semi-luxuries, which only London's wealthiest residents could consume in large quantities or on a daily basis. Nevertheless, the Old Bailey Proceedings suggest that beef and mutton were widely consumed in Georgian London, in one form or another. In March 1778, for instance, a woman who mended china in Farringdon celebrated her birthday by sharing a leg of mutton with a cabinet-maker and his wife; in 1789, a chair-maker in Moorfields was robbed of 5 lbs of beef valued at 12d; in 1806, a master sawyer from Moorfields ate boiled beef brisket and soup on the Thursday before Christmas; and in May 1812, a newlywed Spitalfields tailor ate roast beef at a public house with his father-in-law, a bricklayer and plasterer.[89] It is important to recognise that a broad spectrum of Londoners paid into and benefited from the Smithfield trade because this must have influenced how droves were viewed and treated in the streets.

For some, interest in livestock was limited to a gastronomic appreciation of meat but for many Londoners it went much further. As we saw in the last chapter, considerable numbers of middling and plebeian

Londoners engaged in animal husbandry in this period, but the city also generated demand for agricultural news, books, portraits and exhibitions. Smithfield itself was a problematic stage – its filth, noise and danger deterred most respectable visitors – but curiosity at home and abroad led to the production of topographical prints and guide books which portrayed a pleasingly 'refined market'.[90] In *Sketches of the State of the Useful Arts… Or the Practical Tourist* (1833), the Rhode Island textile manufacturer Zachariah Allen observed:

> Here you behold collected in a little square of about two acres in extent, the beef which is required for the supply of a numerous population. So closely are the droves wedged together, side by side, that their red backs and white horns appear like the surface of an agitated pond, ever undulating and in motion; and a person apparently might walk over their backs as over a pavement. You may suppose with truth that you have before you an area of an acre of solid beef.[91]

This was the same awe-inspiring scene projected by Augustus Charles Pugin and Thomas Rowlandson in an aquatint engraving published in Ackermann's popular *Views of London* in 1811 (see Figure 23). Their bird's-eye view of the market depicts a hive of commercial activity brimming with livestock, salesmen, butchers and admiring spectators.

Smithfield provided dramatic evidence of the scale of metropolitan meat consumption but also London's astonishing reach over the nation's natural resources. In the 1790s, an agricultural survey marvelled that the fattened cattle of Galloway in south-west Scotland were driven 560 km 'to supply the amazing consumption of the capital'. The animals began this journey at four or five years of age, when they were driven to fairs in Norfolk and Suffolk in time to be fed up in the turnip season. They were then driven down to London where they were 'readily sold' at 'high prices' in Smithfield, 'they being such nice cutters-up, owing to laying the fat upon the most valuable parts'.[92] The exhibition of new and improved breeds attested to the nation's ingenuity and so Smithfield became a showcase of modernity, a living gallery of agricultural progress. In the second half of the eighteenth century, new methods of irrigation, fertilisation, crop-rotation and stockbreeding helped to transform cattle and

23 John Bluck after Augustus Charles Pugin and Thomas Rowlandson, *A Bird's Eye View of Smithfield Market taken from the Bear & Ragged Staff* (aquatint, 1811), published in Rudolph Ackermann's *Views of London* (1811–22).

sheep in Britain. Improved livestock were major achievements of the agricultural revolution but their meat was also an important example of product innovation. It has been estimated that carcass weights increased by about one-fifth but equally important improvements were made in meat yield and the distribution of fat in different cuts of meat.[93] These changes took place in the countryside but London played a crucial role in promoting, evaluating and to some extent directing progress. The quality of the nation's cattle and sheep was judged, above all, on the appeal of their meat to metropolitan consumers, and breeds which failed to impress were soon ousted from the market square. In the 1720s, Daniel Defoe noted that the black cattle of West Devon were 'fattened fit for Smithfield Market' and sold to 'the Londoners, who have not so good Beef from any other Part of the Kingdom'. Forty years later, a magazine correspondent asserted:

> It is well known, that this metropolis is the great mart of the British empire; whatever is good, whatever is rare, is brought here as to a certain and good market. The best oxen which our grazing counties produce, are always reserved for the consumption of London.

The capital's salesmen and butchers were considered such unrivalled experts that in 1795 the Bath and West of England Society refrained from appraising the value and quality of different carcasses because, it conceded, 'the most satisfactory intelligence on these heads may be obtained in Smithfield market'.[94] This authority was celebrated for much of the period but began to attract criticism in the early nineteenth century from defenders of rural communities. After sampling Wiltshire's fine veal and lamb, William Cobbett worried that the 'WEN-DEVILS' of London would hear of it, 'get this meat away' and 'the people of Warminster will never have a bit of good meat again'.[95]

When droves traversed the metropolis, they embodied news from the nation's farms and fed an agricultural knowledge economy which permeated metropolitan society. This was reflected in and promoted by the press, which produced an increasing volume of information about the health of the Smithfield trade. In May 1775, for instance, the *Morning Chronicle* recorded that 'At Smithfield market for cattle yesterday, beef fell in price 4d per stone; veal 4d but sheep and lambs (of which last there were a great number) fetched a higher [price] than on Monday'.[96] And in 1794, the *Whitehall Evening Post* observed that

> Smithfield has been well supplied with fat stock through the winter. The Leicestershire and Buckinghamshire grass Oxen never came off better; and the Essex, Norfolk, and Suffolk turniped beasts never died in higher condition; in consequence, prime Beef has been more reasonable than is generally the case at this time of the year.[97]

From the 1760s, London's newspapers began to pay greater attention to the regulation, improvement and potential relocation of Smithfield Market to a more commodious site. In March 1765, the *London Chronicle* reported that

> Cattle [were] sold on Friday in Smithfield rather cheaper than they did on Monday, owing, it is supposed, in a great measure to the late prosecution

> of several forestallers, the good effects of which are already felt by all ranks of people.

The following month, the *Evening Post* claimed that 'cunning' farmers at the country fairs had refused to sell their animals to Smithfield's agents because 'they proposed coming' to the market in person to 'sell their cattle at a moderate price, and lay open the Jobbers villainy'.[98] From the late 1780s, some titles began publishing tables of agricultural prices in the capital, including those for cattle and sheep, while longer-term studies of the trade appeared in guides to the city and agricultural surveys.[99] John Middleton and Stephen Theodore Janssen provide the most detailed and reliable data for animals sold at Smithfield in the eighteenth century. Janssen compiled his figures in the early 1770s and Middleton in the late 1790s. Middleton was a land surveyor, agricultural writer and a corresponding member of the Board of Agriculture, dividing his time between his farm in Merton and a residence in Lambeth. Janssen, by contrast, had been a director of the East India Company and a lord mayor of London.[100] The efforts taken by these very different men emphasises that interest in the livestock trade was as much mercantile and civic as it was agricultural.

It would be tempting to assume that public interest was restricted to those involved in the trade but this was not the case. The polite Bloomsbury spinster, Gertrude Savile (1697–1758) epitomised the kind of 'well-to-do' town-dweller who, according to Keith Thomas, should have been 'inclined to think of animals as pets rather than working livestock'. And yet Savile's journal entries in the 1740s and 1750s show that she was acutely aware of, and concerned by, the spread of cattle plague across the country. In March 1746, she lamented that 'The mortality amongst Cows [is] continuing. Any cows or cow calves are forbid to be kill'd (for eating), for 4 Years from Ladyday next.'[101] It would have been difficult for anyone living in the city to ignore the unfolding crisis because London was on the front line. In February the following year, the *Westminster Journal* reported that

> a great Number of Farmers from different Counties, who have been Sufferers by the Distemper among the Horned Cattle, attended at the Office in Hatton Garden, in order to receive the sum of 40s for each of

> them knock'd on the Head, to prevent the spreading of the said contagious Distemper.[102]

The disease is thought to have broken out in the coastal marshes of Essex but the region was a key supplier of veal calves to Smithfield. As well as hosting the most concentrated convergence of livestock in the country, it was from London that cattle plague spread to most of the rest of the country.[103] Three years later, Savile wrote: 'The distemper among the cattle still reigns violently, and is got again about London… A new Order of Council… forbid[s] the driving any Cows, or Calves above 2 miles after the 14th of next month.'[104] Considering her sustained interest, it is perhaps surprising that Savile never referred to the hundreds of animals which were driven past her Bloomsbury home every week, and this reflects a broader pattern in the journals and correspondence of other polite Londoners. Perhaps encounters with livestock were considered inappropriate subject matter for this kind of writing but there were other occasions when such discourse was encouraged.

Meetings of the Sublime Society of Beefsteaks (1735–1870) and other dining clubs celebrated British beef in raucous songs, ceremonies and uniforms.[105] Their membership was exclusive but many more Londoners attended exhibitions of gigantic oxen such as the Royal Lincolnshire Ox, which was sold at Tattersall's horse repository for 185 guineas in May 1790. Advertised as 'the largest and fattest ever seen', it was displayed at the Exeter 'Change menagerie on the Strand for almost a year, fetching 1s per view before being slaughtered. A decade later, the Durham Ox, a beast said to weigh 3,204 lbs and to stand five feet six inches tall, raised nearly £100 in admissions in a single day.[106] Even more impressive were the annual livestock shows held by the Smithfield Club from 1798. The first five shows took place at Wootton's Livery Stables near Smithfield; in 1805, the show moved to Dixon's horse repository in the Barbican and the following year it expanded into the yard of Sadler's Wells in Clerkenwell, where it remained until 1838.[107]

Smithfield ceased to be a livestock market on 11 June 1855. Two days later, the trade resumed at Islington's Copenhagen Fields, at which point it became a suburban operation for the first time in more than 200 years. This was a victory for enlightened urbanity but also reflected wider changes in London's meat supply systems. The advent of haulage by steam locomotives in the 1830s began to erode the proportion of meat supplied on the hoof so that by the early 1850s, three-quarters of what was sold at Newgate Market came from animals slaughtered outside of London.[108] It was, however, Smithfield's closure, just two years after recording its highest ever annual sales, which most dramatically transformed the city's relationship with the animals it consumed. Keith Thomas assumed that Londoners had become 'remote from the agricultural process' well before 1800 but it was only after the trade's removal from inner-city London in 1855 that livestock droves began to fade from daily life for most of the population.[109] Throughout the Georgian period, the Smithfield trade reinforced the connection between the production and consumption of meat, as well as London's identity as an agropolis.

5

Consuming horses

London played a dominant role in British consumer culture in the Georgian period. The capital not only concentrated the purchasing power of the nation's wealthiest consumers, it was also the mainspring and principal shop window of aesthetic taste, fashion and innovation.[1] Historians have written voluminously on the subject from different perspectives but horses make only the most fleeting of appearances. Scholars of the consumer revolution have tended to focus on two categories of goods: exotic imports such as tea, coffee, sugar, spices, silks and calicoes; and manufactured items including clothes, chinaware, furniture, clocks, jewellery, wallpaper, books, cooking utensils and fine art.[2] Meanwhile, historians of shopping have concentrated on the marketing practices of the kind of refined retail outlets which sold these inanimate goods, in the process overlooking or actively side-lining alternative sites of consumption.[3] There appears to have been an assumption that the dung-bespattered stables, repositories, markets and fairs in which horses were traded detracted from, rather than contributed to, polite urban culture. Yet, as this chapter will show, London's elite riding and carriage horses were in the thick of the consumer revolution. These animals were highly sought-after luxury items and as such supported a major sector of the metropolitan economy, a sector that developed innovative marketing practices in this period. But more than this, they were voracious consumers in their own right, compelling their owners to spend huge sums on food, architecture, manufactured goods, medical treatment and human labour. A key aim of this chapter is, therefore, to integrate equine agency into debates about the consumer revolution. In doing so, I will argue that consumption of and by horses fostered a pervasive knowledge economy, a sophisticated horse sense, which shaped the city's culture and social relations in important ways.

The horse trade in eighteenth-century England involved thousands of stud farms, breeding grounds, fairs, markets, racecourses, stables and repositories dispersed across the country, but within this infrastructure, nowhere was more influential than London. As the tenth Earl of Pembroke advised his twenty-one-year-old son in 1781, London was the best place to buy horses, better even than Yorkshire, because he would 'find much greater & better choice' in the capital's repositories; 'Dimmock has a great many', he promised, 'Dawson too, & also a man in St Giles. The Borough too is generally well stocked.'[4] The farmer and writer, John Lawrence, expressed a similar view in 1796, noting that these venues were 'the best markets in the world for brood mares'; and by 1829, he was even more complimentary, describing the metropolis as 'that vast menagerie of horses... a universal mart, to which recourse is had from the extremities of the kingdom, for both the purchase and sale of horses'.[5] The capital catered for all specifications of horse, from worn out hackneys fit only for the knacker's yard to the most thoroughbred of racehorses, but the city excelled at the top end of the market for riding, hunting and carriage horses. London became a magnet for horseflesh in the eighteenth century: the city's dealers scoured the breeding grounds of Yorkshire, the North East, the Midlands and the Home Counties for the finest specimens and boasted in newspaper adverts of filling their commodious repositories with horses 'just come from' or 'Fresh from the Breeders in the North'.[6] Some of these animals were the progeny of Arabian horses of superior bloodlines which aristocratic equestrians had been importing since the mid-seventeenth century. In 1831, William Youatt noted that between 1650 and 1750 a 'system of improvement was zealously pursued' and not only among racehorses: 'By a judicious admixture and proportion' of 'every variety of Eastern blood', he observed, 'we have rendered our hunters, our hackneys, our coach, nay even our cart horses, much stronger, more active, and more enduring'.[7]

The full scale of the metropolitan horse trade is difficult to pin down, partly because it was largely unregulated but also because so many dealers were involved, about whom limited evidence survives. Newspaper adverts and guides to the city do, however, suggest that by the early nineteenth century, around 25,000 horses were sold each year. Based on the conservative estimate that the average price paid was £5, this trade was worth £125,000. Although much smaller than London's gargantuan trade in meat on the hoof, horse dealing dwarfed the trade in many other luxury and semi-luxury commodities. Consider, for instance, that in a twelve-month period in the 1780s, the value of goods sold by the East India Company at home included chinaware worth £24,780; coffee and drugs valued at £70,120; and saltpetre and redwood amounting to £101,400. Only the EIC's trade in Bengal raw silk (£221,890), Chinese silk (£304,800), Bengal cotton piece goods (£987,010) and tea (£2,202,520) were significantly more valuable than London's trade in horses.[8]

In the early eighteenth century, business was dominated by Smithfield's Friday market for 'ordinary horses' – a loose category comprising saddle, carriage and cart horses – which took place soon after the departure of the cattle, sheep and pigs.[9] Smithfield had hosted horse trading from at least the late twelfth century but its operations ballooned in the early modern period.[10] In the mid-1640s, Smithfield dealers became powerful contractors to the parliamentary armies, supplying thousands of horses each year; and in the early 1700s, Daniel Defoe described the market as a place where 'great numbers of horses, and those of the highest price, are to be sold weekly'.[11] No official records of the scale or value of the market appear to have survived for the eighteenth century but some estimates were made at the end of the period. In 1828, Smithfield's inspector of police estimated that 300 or 400 horses were generally brought to market each week, which gives an annual total of 15,600–20,800.[12] Even if we allow for a degree of exaggeration, this remains a significant trade and yet its reputation rests largely on commentary associated with the campaign to end the marketing of live animals in the area. Between 1825 and 1855, certain guides to the metropolis wrote disparagingly about the quality of horses on offer at Smithfield: in 1851, the market was said to be 'more noted for knackers than for high mettled racers' and a place where 'low jockeys attempt to display their broken-down animals to the best advantage'.[13] These observations were not entirely new: the late eighteenth century produced caricatures such as *The Bargain – A Specimen of Smithfield Eloquence* (1780), which features a dealer giving a long-winded sales pitch about a decrepit horse (see Figure 24). Such depictions were infused with equestrian snobbery but were generally light-hearted; by contrast, in the 1820s, the market began to face more serious criticism for alleged cruelty, nuisance and criminality. In 1828, the inspector complained that the dealers were 'the most lawless set I ever saw', not least for running their horses up and down the streets, behaviour which gave the market the sobriquet: 'Smithfield Races'.[14] The following year, John Lawrence described the market as the 'epitome of hell' on account of the 'miserable objects destined for slaughter' which were sold there.[15] But not all depictions from this period were negative. *Old Smithfield Market*, a painting completed in 1824 by the Swiss artist Jacques-Laurent Agasse, presents a bustling scene featuring respectably dressed buyers inspecting physically impressive beasts (see Figure 25). The artist offers a selective view of proceedings but he studied the metropolitan horse trade for more than forty years and was viewed as an expert by grooms, dealers, and horse fanciers alike. Thus, this image supports the impression that horse dealing continued to thrive in Smithfield throughout the Georgian period.[16]

In the second half of the eighteenth century, however, the horse trade became increasingly fragmented and Smithfield conceded market share to stables and repositories dispersed across the metropolis. Selling horses from stables was nothing new but the scale and sophistication of the activity which developed in this period was unprecedented. In the 1740s, around the same time that metropolitan inns started to hold auctions of household property and other goods, the livery stables attached to these inns started selling multiple horses on a regular basis.[17] In 1754, the owner of Bishopsgate's Four Swans Inn informed readers of the *Public*

24 Anon., *The Bargain – A Specimen of Smithfield Eloquence* (etching and engraving, 1780).

25 Jacques-Laurent Agasse, *Old Smithfield Market* (oil on canvas, 1824).

Advertiser that he had 'fitted up a compleat Room for twelve Horses' to stand prior to sale and that auctions would take place every three weeks throughout the year.[18] The practice of advertising horses for sale in the London press began around 1703: the earliest example I have found appears in the *Daily Courant* and concerns 'an extraordinary well condition'd Chaise Gelding' which was to be sold for £10 at the Cosar's Head in Westminster.[19] Initially, such adverts only appeared sporadically, very few stables featured more than once and the vast majority of sales involved a single horse. This all changed in the second half of the century, at a time when there was an explosion in the number of titles, issues and pages being published. In the years 1703–19, there were generally seven newspapers in circulation; by the early 1750s, this had doubled; and by 1780, there were seventeen titles, each offering far more advertising space than ever before. The first half of the century witnessed a dramatic increase in the number of stables advertising horse sales in newspapers. Searching the Burney Collection online, I found sixty different sites advertising in the first two months of 1752 alone, more than I could find in first twenty years of the century combined. In 1757, however, the duty on adverts was doubled to two shillings and this made it uneconomical for stables to advertise small sales. By the early 1780s, the number of stables advertising had slumped by around 80 per cent as a few elite repositories began to monopolise the practice to attract crowds of bidders to sales of multiple lots.[20]

London's first horse repository was established around 1740 in Little St Martin's Lane, a commercially advantageous location at the heart of the West End between the King's Mews and the city's main coach-building district. The firm's proprietor, Nathaniel Bever, began advertising in 1753 and as competition intensified, his successor, James Aldridge, marketed the firm as the Original Repository. There was a fine line between a repository and a large livery stable dealing in horses; generally, the latter conducted private sales of individual horses while repositories held weekly auctions in which multiple lots went under the hammer. By the 1830s, Bever had been emulated by at least fifteen other dealers and the auction days of the leading repositories were common knowledge: Mondays belonged to Tattersall and Wednesdays to Aldridge; while Tuesdays and Fridays were shared by Dixon at the Barbican and Sadlers of Goswell Street.[21] By the late eighteenth century, repositories dominated the trade in high-quality saddle and coach horses. Writing in the 1820s, John Lawrence found them to be 'beyond a doubt, the best adapted to the disposal of horses of high qualification, and for which great prices are expected'. Yet despite this focus on the upper end of the market, these sites continued to stock a range of horses. Tattersall handled the most valuable bloodstock, regularly selling famous racehorses, the finest Arabian stallions and the highest-bred hunters, but he also appreciated the need to cater for smaller budgets by offering more modest hunters, coach horses and hacks. The firm generally drew the line at working horses but 'useful boney Geldings', ideal for 'heavy Draft work' were available at St Martin's Lane and the Barbican.[22] The longevity of the Original Repository (1753–1926), the Barbican Repository (1773–1926) and above all, Tattersall's at Hyde Park Corner and later

Knightsbridge (1766–1939), emphasises that the Georgian period was the golden age of metropolitan horse dealing.

These enterprises commanded commodious sites, often in some of the best parts of town, and held stock worth hundreds if not thousands of pounds at a time. In 1792, Joseph Aldridge insured eight stable blocks and their contents for £2,310 while Edmund Tattersall's stables and coach houses were valued at £3,100 excluding their precious livestock.[23] By then, around 5,000 horses passed through Tattersall's every year and the trade's leading firm had appeared in numerous topographical prints (see Figure 26), caricatures, songs, plays and novels, as well as in sporting and equestrian literature. In the 1820s, George Young's Horse Bazaar in Portman Square, occupied two acres and boasted extensive stabling, attended by an on-site veterinary surgeon; while the

26 Thomas Sunderland after Thomas Rowlandson and August Charles Pugin, *Tattersall's Horse Repository*, coloured aquatint, published in Rudolph Ackermann's *The Microcosm of London* (1809).

London Repository on Gray's Inn Lane Road, was commended for its 'lofty, light, [and] airy... accommodation for about two hundred horses, and galleries for more double that number of carriages'.[24] Historians of the consumer revolution have often admired the giant drapery and haberdashery shops which opened in the late eighteenth and early nineteenth centuries, but London's horse repositories predated these sites and were no less innovative or successful.[25] As well as offering the best choice of horses, repositories sought to provide a pleasurable shopping experience; we know, for instance, that the auction rooms at Aldridge's and the Horse Bazaar were both flanked by seated galleries which facilitated an unimpeded view of the lots while also creating a suitable space for polite conversation, elevated from the hubbub of the sales floor.[26] Particular praise was reserved for the subscription room at Tattersall's which was always open on sale days for wealthy buyers to discuss the merits of the animals going under the hammer and to celebrate purchases.[27]

Riding and carriage horses were expensive commodities. In the 1770s, gentlemen riders deprived of their geldings by thieves generally valued their animals at between £15 and £20 – this was roughly equivalent to the price of a pair of silver candlesticks or a coachman's annual wages – although inferior saddle horses could be bought for as little as £2.[28] Buying a horse for recreation was well within the reach of many middling Londoners, as shown by tax records for the parish of St Leonard's, Streatham, where those paying duty on a riding horse in 1800 included an apothecary, shopkeepers, a publican, a tailor, a cooper and a butcher; while a school master, a shopkeeper, a salesman and a brewer were taxed for a two-wheeled carriage and a horse.[29] Yet, the main costs associated with horse ownership lay not in the purchase of the animal but in its incessant needs for shelter, food and care. In *The London Advisor and Guide* (1786), John Trusler calculated that it cost more than £20 to keep a horse in a London livery stable for thirty-two weeks of the year, roughly the same, by his calculations, as sending a child to boarding school for a year. The hostler demanded £1 1s but the major expense, £19 12s, was generated by food: in just one week, a riding or private coach horse consumed a truss (~ 25 kg) of straw, two and a half trusses (~ 40 kg) of hay and five and a quarter pecks (~ 47 l) of oats.[30] These animals and the equestrian culture which they facilitated were also voracious consumers of manufactured goods: their owners paid handsomely for iron shoes, saddles, stirrups, bits, reins, coach harnesses, whips, combs and horse cloths. In the 1760s and 1770s, stolen saddles were mostly valued at between five and fifteen shillings; and while a standard bridle might only cost one or two shillings, those incorporating a bit plated with silver could cost £2 or more.[31] Altogether, the tackle required to maintain and ride just one horse added up to several pounds, with inevitable wear and tear necessitating additional expenditure on regular repairs and replacements. By far the most expensive of all equestrian items were vehicles: in the late eighteenth century, a basic four-wheeled open carriage of the phaeton type could be bought for £37 but heavier models equipped with optional extras could cost at least ten times as much. The metropolis dominated the nation's lucrative coach-building trade: Pigot's *Directory* for 1827 identified 567 specialist firms, 4 per cent of all the metropolitan manufacturers which it listed. In addition, London's coach and riding horses supported 300 firms involved in the saddle- and harness-making trades.[32] Fifty years earlier, these animals also kept at least thirty-six blacksmiths busy.[33] While iron horseshoes were relatively inexpensive items – a set of four cost around 20d in the 1760s and a horse at work in the city required around eight sets a year – each one had to be fitted by a trained hand.[34] Moreover, during the eighteenth century, this work was subsumed into an expanding array of increasingly sophisticated farriery services.

As Michael MacKay has shown, London was the prime mover in the medicalisation and commercialisation of horse care in Georgian England. Highly prized animals belonging to the West End elite created a booming market for equine medicines and the establishment of several subscription hospitals in the second half of the century. The city's leading farriers or horse doctors, as some came to be described, formulated and marketed a profusion of cordial balls to be consumed after a hard ride or at the first signs of colic and gripe; oils and pastes to apply to wounds, fistulas and tumours; pissing balls, which purged animals with intestinal problems; various 'drinks', medicines in liquid form; and 'drench' or enemas. These treatments became the most prevalent form of horse medicine in this period and their profitability drew surgeons, apothecaries and druggists into the trade.[35] London's wealthiest equestrians required regular visits from farriers and paid eye-watering sums for their services. In the 1770s, the Marquess of Rockingham's horses in Grosvenor Square ran up standard monthly bills of around £7 and more complex treatments could more than treble this.[36] Twenty years later, horses belonging to the third Earl of Egremont in Piccadilly required the services of the shoeing farrier Henry Boulton around forty times a year, as well as less frequent visits from the aristocracy's preferred farrier. Edward Snape opened England's first known horse hospital in Knightsbridge in March 1765 and charged wealthy subscribers for stabling, food and medicine for as long their animals required treatment. Snape's hospital closed in 1778 but others soon followed, the most significant being the London Veterinary College Infirmary – which opened in Camden Town in 1791 with stalls for more than 100 horses – and the equally large equestrian receptacles and medical dispensaries which William Taplin established in Somers Town in 1793 and on the Edgware Road in 1796. In addition to administering medicines, these sites performed surgical operations, mostly to treat lameness.[37] The medicalisation of equine care generated new possibilities for sick and injured horses but also far greater costs for their owners. These were remarkable developments but an even more dramatic expression of equine consumer power was its impact on the built environment.

The Georgian West End is often described as an 'innovation in urban living' but historians rarely cite the construction of thousands of stables and coach houses, laid out in carefully planned mews complexes, as evidence of this.[38] This is a serious oversight because mews were not just an ingenious solution to the stabling needs of London's horse-owning elite, they were crucial to the very success of the town house as a mode of polite urban living, more specifically as a base from which to engage in recreation, sociability, politics and business. London's earliest mews appeared in Covent Garden in the 1630s but the vast majority were built after 1720. In the 1740s, there were still only twenty-nine examples but by the early 1800s, this had increased to 117.[39] The challenge that architects faced in the West End was where to position buildings for horses and vehicles given that the terrace offered neither a forecourt nor space between properties. Their solution was to build a modest two-storey structure behind the main residence, connected by a garden or yard but accessed from the street by a separate alley. The basic mews layout featured a stable on one side of the ground floor and a double-doored coach house on the other, while the first floor provided a hayloft and basic living quarters for servants. Crucially, the mews allowed owners to keep their precious animals, vehicles and workmen close-by while isolating the household from their stench and noise.

The Grosvenor estate in Mayfair was the first to roll out large-scale mews provision: in the 1720s, the surveyor Thomas Barlow set about transforming its 100 acres into an ordered grid of wide thoroughfares which gave horse-drawn carriages greater manoeuvrability while also creating space for mews behind some of the terraces.[40] One of the largest mews units on the estate belonged to the second Marquess of Rockingham, at No. 4 Grosvenor Square. In residence from 1751 to 1782, Rockingham was a leading horse breeder, racehorse owner, huntsman and patron of George Stubbs. At Wentworth Woodhouse in Yorkshire, he built one of Europe's grandest stables to house 100 horses and in London, the marquess gave his animals the best care the city could offer. According to an inventory from 1782, No. 4 benefited from 'roomy Stabling' for twenty-four horses and 'Standing for four Carriages' in Three King's Yard. These arrangements were second only to the King's Mews at Charing Cross, but most aristocratic residents invested heavily in stabling.[41] In the 1770s, Robert Adam constructed a fourteen-stall stable and generous double coach house to serve Sir Watkin Williams-Wynn's lavish residence at No. 20 St James's Square. Taking the harness room and accommodation for servants into account, these facilities swallowed one-fifth of Wynn's entire plot. By today's standards, mews did not offer horses adequate space or ventilation but at the time, their owners were making extraordinary financial and spatial sacrifices to keep their animals nearby and in the best possible condition.[42]

Stabling became an increasingly important factor in the appeal of West End properties, streets and even entire estates. The Grosvenor estate's mews provision remained irregular and in the second half of the eighteenth century this contributed to a gradual shift in appeal towards the newer Portland and Portman estates in Marylebone, which placed the needs of horses at the heart of a more consistent blueprint for fashionable urban living. In the 1720s, there was a coach for every four houses in the parish of St George's, Hanover Square, on the Grosvenor estate.[43] By 1800, newer parts of the West End were approaching the point where virtually every house had a private stable and coach house (see Figure 27).[44] Peter Potter's detailed 1832 *Plan of the Parish of St Marylebone* reveals that eight in ten houses on the Portland estate were served and that virtually every house on Upper Wimpole Street, Upper Harley Street and Devonshire Place, which were built in the final quarter of the century, benefited from private stabling.[45] Elite house-hunters showed great determination to make strategic choices and when they placed newspaper adverts for 'Wanted' properties, stabling ranked highly among their specifications.[46] In the late 1780s, for instance, the *World* newspaper published an enquiry for

> A House to Rent... elegantly furnished or unfurnished, fit for the reception of a large family, with double coach-house, and stabling for not less than four horses; the situation preferred will be the neighbourhood of Cavendish or Portman Square.[47]

Property-for-sale adverts reveal that most mews on the Portland estate offered stalls for five or six horses and standing for two coaches which would suggest that the estate as a whole could accommodate around 1,500 horses.[48] But mews were much more than storage units, they were sophisticated horse servicing zones which employed and housed teams of specialist workers. Closer examination of this half-hidden world reveals that horses were greedy consumers of human labour, as well as space and architecture.

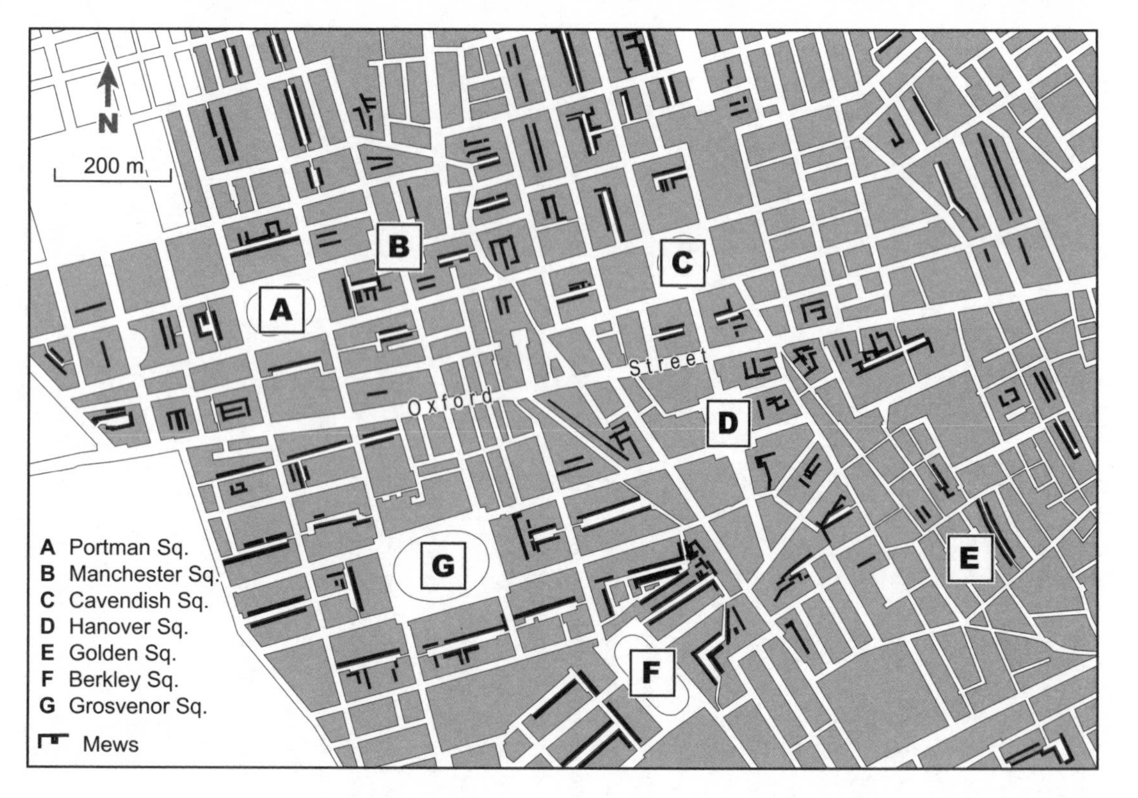

27 Mews in the Grosvenor, Portman and Portland estates *c.* 1799, based on Richard Horwood's *Plan of the Cities of London and Westminster, the Borough of Southwark and Parts Adjoining Showing Every House* (engraving, *c.* 1799).

The West End usually attracts attention for its elite human residents. J. J. Hecht first drew attention to the area's domestic servants in the 1950s and a handful of more insightful studies eventually followed in the 1990s and 2000s. Yet, with the exception of Tim Meldrum, historians have overlooked the thousands of people and horses living in mews in the eighteenth and nineteenth centuries.[49] This is not for want of evidence or because these actors were insignificant in number or influence. Owners of even modest equipages had to employ at least one coachman and one groom, while the city's leading equestrians required much larger retinues: in 1782, the Marquess of Rockingham employed two coachmen, three postilions, three grooms and a stable boy, meaning that two-fifths of his London servants worked with horses. Across an entire estate, therefore, mews workers formed a large occupational group; the Portland estate alone would have needed at least 500 men and the West End as a whole almost certainly employed over 2,000 men by the late eighteenth century. They were poorly paid – in the 1780s, a coachman in London could expect to earn around £16 a year, while grooms received up to £10 – but across the West End, the total value of the human labour consumed by riding and coach horses was considerable and rapidly increasing.[50]

Certain features of equine biology make caring for horses unrelentingly labour-intensive. The animal's caecal digestive system makes it necessary to serve at least three separate feeds a day as well as supplying large volumes of water to help prevent their small stomachs and intestines from twisting and blocking, a condition known as colic.[51] Horses also produce large quantities of dung – in the 1840s, Henry Mayhew estimated that a single animal could defecate 23 kg in just six hours – which had to be regularly cleared from the stable both to keep the animals in pristine condition and to guard against infection.[52] Together with the time-consuming task of grooming, these basic equine demands contributed to the long and unsociable hours which West End stablemen were required to work. Carolyn Steedman has rightly challenged Adam Smith's formulation of the servant's labour as a kind of non-work, or anti-work, as well as E. P. Thompson's suggestion that servants were not part of the working class because they were not really workers.[53] Steedman leaves no doubt that cooks and maids were real workers, or that vegetables and dirty clothes 'all have their wants: they tell the worker what needs doing to them'.[54] The same was certainly true of horses, which generated a gruelling regime of repetitive tasks, but the work of coachmen and grooms went far beyond the basic 'wants' of horses. These individuals were involved in the improvement of equine performance as well as elite mobility and display, a process that gave increasing attention to orderliness, cleanliness, individualised care, punctuality, dietary control and record-keeping.

This responsibility had a major impact on work cultures and social relations, both within and beyond the architectural confines of the mews. To begin with, the round-the-clock demands of horses and their owners meant that coachmen and grooms typically worked longer than the twelve-hour average for metropolitan workers calculated by Hans-Joachim Voth. It also meant that these men usually had to live in basic quarters above the stable and coach house, sometimes with the coachman's family, to fulfil their duties.[55] Late-night sociability forced some coachmen to fetch their masters from clubs, only to rise a few hours later to tend to their horses' needs. A coachman's work generally commenced at five or six o'clock in the morning, which according to commentators in the 1780s, was two or three hours before many domestics had even climbed out of bed.[56] The need to respect the schedules and spur-of-the-moment demands of particularly mobile employers, while also attending to the horses and vehicles, made mews work an exhausting juggling act. This fostered distinctive behaviours and character traits.

Although coachmen and grooms ranked among the household's inferior manservants, astute masters looked for an array of skills and personal qualities to ensure the wellbeing of their valuable animals and the orderliness of their equestrian affairs. Coachmen not only had to ride and drive well but also manage a team of subordinate grooms and postilions; maintain standards of horse care and vehicle maintenance; as well as deal with visiting tradesmen. Meanwhile, a groom had to be 'a complete and perfect master of every part of stable discipline' and to display 'obedience, fidelity, patience, mildness, diligence, humanity, and honesty'. William Taplin warned his wealthy readers that 'the HEALTH, SAFETY, and CONDITION, of every horse' depended upon 'the sobriety, steadiness, and invariable punctuality, of the groom; and by his incessant attention only can they be insured'. Due to their 'arduous task', the 'variegated nature of their servitude, and the property entrusted to their care', Taplin advised that grooms were entitled to 'all the equitable pecuniary compensation, and personal kindness, their employers can possibly bestow'.[57] And working with highly esteemed horses did give West End stablemen a certain amount of power to manipulate master–servant relations and take advantage of opportunities in London's wider equine economy. Tim Meldrum and Carolyn Steedman have rightly challenged Sarah Maza's assertion that eighteenth-century servants lacked autonomy and were caught in a social limbo between their masters and the wider world. Meldrum has argued that most domestic servants 'were engaged in too much interaction with others' both within and outside the household 'for them to be in any way aloof or withdrawn', and suggests that these characteristics were particularly pronounced among coachmen and grooms.[58] While physically enclosed, mews were fully integrated into a thriving equestrian economy: a survey of the Grosvenor estate in the 1790s records that a remarkable 142 householders were involved in horse-drawn transport. They included thirty stable-keepers and twenty-three coach-makers, as well as farriers, wheelwrights, saddlers, horse dealers and coach brokers.[59]

As elite horse care became more sophisticated, mews workers gained more and more contacts – in the mid-nineteenth century, one resident of a large West End mews calculated that '100 different street-traders resorted thither daily' – and developed an increasingly lucrative understanding of London's equestrian consumer culture.[60] This could prove a headache for employers: as early as 1731, Jonathan Swift identified several ways that coachmen and grooms could take advantage of their masters and over the next century, mews workers found countless other opportunities to profit from their privileged access to valuable horses and their ability to tap into the trade in horses, vehicles, tackle and provender.[61] This behaviour was so pervasive that even the King's Mews in Charing Cross struggled to restrain its workforce. In 1769, its clerk recorded that

> several great abuses have been practiced… by some of the Livery and others; such as buying and selling, keeping & letting of Horses, & horses & Chaises; & buying and selling Harness, Carriages &c by which means the Mews has been made a kind of Trading Place to the great Dishonour of the King.[62]

These activities were banned 'upon pain of suspension or discharge' but twenty years later the King's Yeoman Rider complained that it was 'incompatible with my situation… to use the means of increasing my income, which those beneath me… have done, and can do, with Propriety; such as buying and selling Horses, and breaking Horses for Gentlemen'.[63] Across the West End, mews workers sought to capitalise on their expertise and contacts to improve their lot, either by upgrading masters or going self-employed.[64] In the 1720s, William Black served as postilion and second coachman to the Earl of Bristol for three years; then transferred to the mews of Sir William Shirkland and Lord Scarsdale, each for two years; and finally left service to become a hackney coachman and hostler at the Hole in the Wall in St Clement Danes.[65] A generation later, John Jennings rose from being a gentleman's postilion in Pall Mall to become Viscount Howe's coachman, before going to ride post for a stable-keeper in the Haymarket, all in the space of seven years.[66] The lives of mews workers revolved around equine consumption more than any other occupational group but the expansion of the horse trade drew in men from across the social spectrum.

When describing Tattersall's, Pierce Egan's fictional man about town, Corinthian Tom, observed that 'It is no matter who sells or who purchases at this repository'.[67] This was an exaggeration but horse dealers were as diverse as the animals they sold. By the late eighteenth century, there were more than forty individuals who formally identified themselves as horse dealers in wills, insurance policies and in court, but the trade also engaged thousands of people for whom horse dealing was not their principal occupation.[68] Hostlers, coachmen and grooms, men who benefited from convenient access to stabling and insider knowledge, were particularly active but almost anyone could dabble in the trade. In 1824, a bricklayer from the Hornsey Road coyly informed the Old Bailey that he did not consider himself a horse dealer because he had only 'bought four poneys within the last two years'. And in 1795, a Whitechapel shoemaker recalled that he had exchanged a pony and ten guineas for a gelding, adding that he generally kept a horse in summer but sold it in autumn.[69] It soon emerged that the shoemaker's acquisition had been stolen and it was not uncommon for the trade to lure poor Londoners into crime. In 1733, a button-maker from Westminster was found guilty of stealing a brown gelding, which he had brazenly offered for sale in Holborn. Even bolder was the thirteen-year-old boy who lay out in a field in Hackney until three o'clock in the morning in 1775 to steal a mare, which he then offered for sale at Smithfield's Greyhound Inn. The rewards could be great but horse stealing was a capital offence and both thieves were sentenced to death.[70] Plebeian horse dealing and its close associations with criminality made the horse trade a source of considerable anxiety for polite Londoners, as shown by a heated exchange between the barrister William Garrow and Francis Hall, a labourer from Winchmore Hill. Testifying as a witness in a horse theft trial at the Old Bailey in 1784, Hall recalled:

Hall: A month ago last Friday, I was going from London home, through Kingsland turnpike, and I saw the prisoner getting off a mare in the road. I asked him where he was going with her… he said to Smithfield; I asked him if she would draw, and he said yes; I asked him the price of the mare, and he said seven pounds. I said she is in a very bad condition to sell, I think you will not get that money for her…

Garrow: What made you enquire about this mare?

Hall: It is a natural case, when anybody is going to Smithfield, and one sees a horse in sweat and dirt.

Garrow: A natural case. What do you make it your business?… And do you ask every man with a horse in a sweat and dirty, in Kingsland-road, where he is going with him?

Hall: I have asked several.

Hall accepted that he had been looking to buy a horse but denied Garrow's allegation that he sold stolen animals to slaughterers.[71] The case emphasises that a broad spectrum of Georgian men took an active interest in the horse trade but also that it generated tension between different tranches of society. In a trial heard at the Court of King's Bench in 1762, Stephen Gardner explained that he had been a master shoemaker but also dealt in horses and that two years earlier he had sold Joseph Hughes, a prosperous farmer at Walthamstow, a saddle horse. A fortnight later, Hughes complained that the animal was diseased and insisted that he take it back. Gardner refused, arguing that the horse had not been warranted, and the dispute escalated. Gardner was coerced into signing a promissory note to reimburse Hughes and was imprisoned for non-payment. On his release, he claimed that events had reduced him to a journeyman shoemaker and he brought a case to trial in which Hughes was convicted of extortion.[72] Conflict was not restricted to the lower end of the market: in 1787, Lord Herbert, the future Earl of Pembroke, entered into a dispute with Richard Tattersall, who had begun his career as a lowly stud groom. In an indignant letter, Herbert complained:

> As it is a fortnight since I purchased a horse got by Sypher for seventy five guineas at your auction, which horse turns out to be unsound, & it is nine days since you took him back, & that I have called twice at your house, & have not heard from you, you cannot reasonably expect me to wait longer. I have therefore given a draft upon you, payable to the bearer, for the purchase money. The idea of the horse not being unsound is too ridiculous & absurd.[73]

Horses made for risky acquisitions in this period: appearances could be deceptive and an animal's value could change rapidly and irreversibly. Consumers did receive some protection from the law, which directed that 'if a Man sells a Horse, and warrants him to be Sound of his Wind and Limbs, if he be not, an *Action upon the Case* lies'.[74] Horses advertised for sale in the London press were often 'warranted sound and free from vice', meaning that the seller guaranteed them not to be lame or unmanageable, but these agreements were opaque and legally unstable.[75] Sellers accused of breaking warranty could easily claim that any lameness had occurred after the sale or that poor riding was responsible for an animal's unruliness. By the 1820s, it was generally accepted that warranted horses had to be returned within two days of sale 'on alleged failure of the warranty' but this did not eradicate disputes, as lameness and other health problems could take much longer to emerge, while serious diseases such as glanders and problems with an animal's eyesight were not consistently warranted against. Even at the top end of the market, buyers were left perilously exposed. In the city's repositories, auctioneers commonly prefaced sales with the words: 'I am instructed by the proprietor, to say this horse is sound, but to avoid trouble, he does not choose to warrant him'; while some vendors offered to sell a horse warranted at one price or unwarranted at a substantial discount.[76] Throughout the period, therefore, all buyers were forced to uphold the principle of *caveat emptor* ('let the buyer beware') and hone strategies to manage risk. The behaviour of a hostler in Grosvenor Place in 1810 provides a shining example of this: when a vendor refused to warrant a gelding sound because 'he did not want to have further trouble with him', the hostler remained calm, 'had the horse turned up and down the yard, to see if he was sound' and, reassured, agreed to pay fourteen guineas.[77]

This labyrinthine department of London's consumer culture inspired the publication of *The Adventures of a Gentleman in Search of a Horse, by Caveat Emptor*, in 1835, a feast of a book offering satirical tales, practical advice, a summary of pertinent legal cases and a list of conditions of sale at the leading London repositories. Its author, 'a lawyer by profession, and a jockey by taste', projected a beguiling equestrian world and urged newcomers to develop a head for horses to avoid losing their way.[78] For the novice equestrian, Sylas Neville, the task was at first daunting and then frustrating. In the first half of April 1769, he tried out 'a little bay stone-horse', inspected 'two small stone-horses' at livery in the City and considered a 'bay gelding, much recommended by Bever for soundness, gentleness etc'. He was soon exasperated and conceded, 'it is very difficult to find a good horse'. The first, he complained, 'has not foot enough to carry me a hunting, was always vicious & is of late become more so', while the stone horses 'have not full tails' and Bever's was the wrong colour.[79]

As with other luxury goods sold in the metropolis, the horse trade was highly sensitive to the fluid aesthetic preferences of a fickle West End elite but buyers could rely on certain principles. It was widely accepted, for instance, that coach horses should 'match well in height, in form, and in colour; they should step and pull well together, and their trot should be as equal as possible'.[80] This was important because to ride in a carriage drawn by identical or symmetrical horses not only conveyed good taste, it also demonstrated the owner's ability to maintain an orderly equipage and, by extension, an orderly household. Equestrian consumers became increasingly demanding in the eighteenth century, prompting sellers to provide detailed descriptions of their stock and to match compatible animals. As well as recording an animal's colour, adverts often recommended them for having a 'star' or 'blaze' on their foreheads, a 'snip' on their nose, long or cropped tails; and white feet, arranged in a fore or hind pair. Some buyers became overly fastidious, behaviour which made the pragmatic John Lawrence complain that 'The match of colour is surely of the least consequence, and a good pair of horses should not be rejected for a few shades of variation'. All but the most ignorant of buyers recognised the importance of strength, stamina and temperament but even here, preferences changed. Lawrence observed that it had once been the fashion 'to drive mares' which led to 'a peculiar class of strong, short legged, bold and high crested' animals known as gig mares being bred and sold at a high price, but 'They had their day and it has since been decided by our knowing ones, that the gelding is to be preferred to the mare, for his superior steadiness in harness'.[81] Horse dealers took advantage of these complexities and became infamous for their puff and patter. In newspaper adverts, they promised that animals had 'a great deal of blood' and 'good Meat in their Belly', or that they were 'clever', 'nearly thoroughbreds', 'good goers', 'remarkably fast movers' and fine 'in all their paces'. This culture provided rich fodder for satirists: William Bunbury, who was highly literate in equine matters, quipped in 1787 that

> As a purchaser, it is immaterial whether you go to Tattersall's or Aldridge's, to Meynell's Hunt, or his Majesty's, it is probable you will be taken in

> wherever you go. To define a perfect horse is nearly impossible, and to tell you where to buy one, completely so.[82]

In addition to aesthetic concerns, buyers also had to look out for a catalogue of diseases and injuries, the detection of which was complicated by dealers, farriers and stablemen using French and Latin terms, as well as their own jargon. Between them, Francis Grose's *Classical Dictionary of the Vulgar Tongue* (1785) and James Caulfield's *Blackguardiana* (1793) recorded that 'Bone-setter' meant a hard-trotting horse, 'piper' a broken-winded one and 'rip' a 'poor lean worn out horse', while 'rum prancer' was the cant term for a fine horse.[83] Bunbury's *Academy for Grown Horsemen* parodied a popular strand of equestrian literature which promised to arm gentlemen against unscrupulous dealers and tradesmen. From the 1730s, various self-help books began to encourage wealthy equestrians to learn about farriery so they could supervise the treatment of their animals, but the inclusion of symptoms and anatomical information was equally valuable to nervous buyers. These guides absorbed major advances in equine medicine in the late eighteenth and early nineteenth centuries, culminating in works by William Taplin in 1788 and 1796, and William Youatt's *The Horse* in 1831.[84] Another revealing guide from this period was *Ten Minutes Advice to Every Gentleman Going to Purchase a Horse Out of a Dealer, Jockey, or, Groom's Stables* (1774), which began by complaining that 'deceptions' were seen 'in a less fraudulent light than they seem to deserve' before offering tips for spotting diseases such as glanders, strangles and morfoundering, as well as the signs of a 'moon-eyed horse'. Its author sought to expose the techniques used by some dealers to disguise signs of ageing, injury and disease. These included rasping and blacking hooves to hide circled feet, scorching limbs to remove blemishes and filing or burning ('bishopping') teeth to make a horse appear less mature.[85] Youatt lamented that 'some of the lower class of horse-dealers' concealed sunken eyes, a telltale sign of old age or illness, by puncturing the skin and

> with a tobacco pipe or small tube, blow into the orifice, until the depression is almost filled up. This operation is vulgarly called *puffing the glims*,

> and, with the aid of a bishopped tooth, will give a false appearance of youth, that will remain during many hours, and may deceive the unwary, though the puffing may easily be detected by pressing on the part.[86]

There was even a technique to make a sullen horse appear 'lively and carry his tail well': *Blackguardiana* explained that to 'feague a horse' meant inserting ginger up its 'fundament'.[87]

All of this emphasises that horses gave low-born men an opportunity to use experience and equine literacy to profit from the ignorance of wealthier individuals. Many dealers did, however, offer 'a reasonable trial' period which enabled prospective buyers to ride a horse for a few hours before making a decision.[88] John Lawrence viewed this as essential and warned that mistakes were usually made when purchasers arrived just in time for the sale and 'their spirits being exalted, and their eagerness whetted by the eloquence of the orator, the flourish of the hammer, and the crack of the whip, they dash at an extempore bargain'.[89] Caricatures depicted buyers prudently peering into the jaws of horses to inspect their teeth (see Figure 24) and the anonymous author of *How to Live in London* (1828) warned his 'worthy, unsuspecting, self-sufficient' readers to 'look before you leap' to avoid purchasing

> an animal afflicted with spasm, speedy-cut, wind gall, corns, broken knees, staggers, gravel, and cancer in the tongue; a roarer that has been eating hay chops, that has been blistered in the knees, fired in the hock, or (if deficient in these points) one that has been *stolen*.[90]

The Old Bailey Proceedings reveal that thieves brought stolen horses from as far afield as Leicestershire and Somerset to sell them at the highest possible price in London.[91] The illegal trade relied on hostlers, coachmen and grooms to act as accomplices, or at least to turn a blind eye. Henry Mayhew observed that after stealing a horse, thieves would generally hide the animal in a stable or outhouse until the 'hue and cry' had subsided and that some 'low' horse dealers would assist thieves by agreeing to sell stolen animals at Smithfield or at markets outside the metropolis.[92] Certain inns and livery stables appear in the Proceedings more regularly than others, suggesting that they had become safe havens. Sites such as the Greyhound in Smithfield, the Pewter Platter on St John's Street and the Black Horse in Aldersgate benefited from their proximity to the horse market, while evading its toll-book and prying eyes.[93] We can glean a sense of this involvement from an article in the *World* newspaper in June 1788. When Lord Bayham discovered that a horse had been stolen from his Kent estate, he immediately dispatched his groom to London, where he found the animal on sale at an inn near Smithfield. The groom apprehended the thief, a scuffle ensued and when the former 'called out to the servants of the livery-stable' for help they remained 'unconcerned spectators'. Despite this, the groom prevailed and dragged the thief to a magistrate. Other newspaper reports and Old Bailey testimonies reveal that some stable workers traded in stolen animals and became trusted middlemen.[94] Some were prepared to remove distinguishing features from stolen horses prior to sale meaning that distressed victims recovered their animals with ears disfigured and tails cut.[95] Organised horse thieves were notoriously difficult to catch and exposed the limitations of eighteenth-century policing. In 1789, the *World* lamented that the crime had 'risen of late to a very alarming rate' and advised that no one should buy a horse from a stranger. The previous year, the activities of a 'numerous and formidable' gang came to light when the son of one of its members led a constable to a stable near Chick Lane where he had seen six or more stolen horses at a time. The boy admitted that 'these horses were commonly sold in Smithfield' and to seeing the gang steal from Hackney and 'divers other places'.[96]

Victims hoping to retrieve their animals depended on London's horse sense. Both Smithfield Market and the city's repositories developed effective strategies to detect and prevent the sale of stolen horses. By the sixteenth century, the law required that a note was made of all horses sold in fairs and markets in England and that the vendors were made known to the toll-taker to account for the sale. In Smithfield, detailed descriptions of transactions were recorded in the toll-book throughout the Georgian period and this information could prove decisive in court.[97] By the 1730s, it was also common for buyers to ask sellers to present a voucher to prove that they owned the animal or had permission to sell it.[98] By then, Smithfield was becoming a nerve centre for information about horses stolen in every part of England. The market's clerks kept a record of animals that had been reported or advertised as stolen, and when sellers failed to present vouchers, cautious buyers were advised to ask the clerk to consult this record before they purchased. The market also proved an ideal place to raise a hue and cry, partly because horse thieves gravitated to it, but also because so many trained eyes were concentrated there. Thus, when William Terry, a hackney coachman, grew suspicious of a gelding on sale at the Swan Inn on Tyburn Road in 1767, he seized the animal, had the thief 'cried in Smithfield market' and displayed a notice on his premises. As a result, the victim reclaimed his horse and the thief was arrested.[99] Metropolitan newspapers played an increasingly important role in this struggle; from the 1770s, the Bow Street office and victims from across the country published adverts requesting information and the return of their animals.[100]

In this climate, repositories made strenuous efforts to guard against suggestions of impropriety to retain the trust of elite clients. It was common practice for proprietors to interrogate sellers about the age, character and health of their horse, as well as asking how they came into their possession. Some if not all repositories kept a record of horses advertised as stolen and usually held animals in livery for a day or more prior to sale, partly to expose faults and disease, but also to facilitate the detection of stolen goods. In 1754, Neville Henson of Bishopsgate's Four Swans Inn stipulated that he would not put a horse up for sale 'unless they have been in seven Days in order to be view'd', and in 1774, one of Bever's servants told a vendor that his 'master would never permit a horse to be taken in and sold the same day'.[101] Nevertheless, the Old Bailey Proceedings reveal that some thieves did manage to sell stolen horses at repositories. In June 1778, for instance, James Durham successfully sold a gelding stolen from Deptford at Tattersall's. Durham was eventually arrested but this owed more to the power of London's equestrian chatter than to effective policing: news of the sale reached the owner by word of mouth during a visit to Smithfield and Durham was apprehended as he returned to collect his money. While repository keepers acquired more respectability than most horse dealers, they were never far above suspicion. In 1780, Thomas Aldridge was arrested but later cleared of aiding and abetting a horse thief who had entrusted him with a mare stolen from Greenwich; and in 1785, a dealer who had sold 'a great number of horses' at Langhorn's repository was convicted of stealing a horse and sentenced to death. In court, Langhorn's son gave the accused a good character, explaining that he had purchased several horses 'in the fair way of dealing' at the repository, but this failed to sway the jury.[102]

There can be no doubt that riding and carriage horses were a divisive force in metropolitan culture. They aroused envy, lured some into criminality and created fertile ground for dispute. At the same time, however, these animals gave Londoners from all ranks and occupations a powerful shared interest. In *Life in London* (1821), Corinthian Tom advises his friend from the country 'if you have any desire to witness "real life"... and to view the favourite *hobbies* of mankind', Tattersall's

> is the resort of the *pinks*, of the swells ... the *dashing* heroes of the military, the fox-hunting clericals, sprigs of nobility, stylish coachmen, smart guards, saucy butchers, natty grooms, tidy helpers, knowing horse dealers, *betting* publicans, neat jockies, sporting men of all descriptions, and the picture is finished by numbers of real gentlemen.[103]

It is tempting to imagine that the lore of animals belongs to the countryside but we have seen that a broad spectrum of Georgian Londoners developed a highly tuned horse sense. This chapter has also shown that riding and coach horses were eye-wateringly expensive to own, which makes the fact that their numbers soared in this period all the more remarkable. By 1754, the metropolis housed 4,255 of England's 9,000 four-wheeled private carriages, plus a further 2,909 two-wheelers.[104] If we make the conservative estimate that two horses were kept, on average, for every four-wheeler and one horse for every two-wheeler, then mid-eighteenth-century London would have housed around 11,400 private carriage horses. And this figure would have been matched, if not exceeded, by the number of riding horses. If we then consider that the number of mews in the city quadrupled between 1740 and 1810, it seems reasonable to assume that by the later date, this convergence of horses had risen to well over 30,000.[105] The next chapter seeks to explain why so many Londoners fell under the spell of these animals and considers their role in metropolitan recreational life.

6

Horsing around

England succumbed to an equestrian craze in the eighteenth century, an irresistible urge to be diverted by horses. Every part of the country contributed something: the finest riding, coach and carriage horses were bred in the Midlands and the north; champion racehorses grazed on aristocratic estates from Petworth in West Sussex to Wentworth in Yorkshire; and either hunt meets or racecourses sprang up around most large towns and cities.[1] Yet, nowhere was more active or influential in this cultural phenomenon than London. It is surprising, therefore, that only one aspect of the city's involvement has received prior attention: the circus.

It is relatively well known that London produced the first and largest circus venues, the most celebrated performers and the finest studs of performing horses anywhere in Europe. The genre first evolved from trick-riding in suburban inn-yards and tea-gardens in the 1760s. These remained itinerant operations until May 1769, when Philip Astley erected sheds and fencing to create a permanent venue near Westminster Bridge Road, for which he sold tickets and advertised. Trick-riding remained the backbone of Astley's show but he introduced dramatic performances in the early 1770s, at which point Charles Hughes established a rival venue south of Blackfriars Bridge. The pair entered into a bitter commercial rivalry, each spurring the other onto increasingly ambitious equestrian spectacles. Astley roofed his venue in the late 1770s and rebranded it as an amphitheatre while in 1782, Hughes went into partnership with the dramatist and songwriter Charles Dibdin to establish the Royal Circus, a large premises with a proscenium stage, at St George's Circus. Both companies embraced literary-based equestrian dramas and by the early 1800s, each could attract more than 2,000 spectators a night.[2] This success posed a challenge to London's traditional patent theatres and in 1811 they gave in to 'hippo-mania'. In February, twenty riders on horseback streamed onto the Covent Garden stage in a performance of *Blue Beard*, generating huge profits which convinced the theatre to stage a play written specifically for Astley's horses. *Timour the Tartar* proved an even greater crowd-pleaser and the craze spread to other theatres.[3] The Haymarket's satirical piece, *The Quadrupeds of Quedlinburgh*, opened with the words 'Dear Johnny Bull… / Your taste, recovered half from foreign quacks, / Takes airings now on English horses' backs'.[4]

Historians have offered various explanations for the rise of the circus in Georgian London but tend to imagine a theatrical genre disconnected from other forms of equestrian culture. By contrast, this chapter contends that the circus appealed precisely because so many Londoners either rode for pleasure or worked with horses and so fully appreciated the skill and sensations involved. The possibility that London fostered a distinctive equestrian culture has been overlooked partly because previous studies have tended to associate horse riding, racing and hunting with the countryside. The field has been labelled 'the great rural diversion', while hunting is generally viewed as a way for rural landowners to distinguish themselves from 'the urban and mercantile, the sedentary and the professional'. Peter Borsay complicated this by showing that provincial towns were crucial service centres for equestrian sport but stopped short of examining the motivations and behaviours of urban riders.[5] More recent studies have acknowledged the role played by the equipage in elite urban lifestyles in this period but have tended to focus on two applications of the horse-drawn carriage: mobility and the display of status.[6]

By contrast, my research reveals that riding was an important mode of sociability and an alluring diversion in its own right. Part of the problem with the existing historiography lies in the assumption that England's landed classes divided their recreational life neatly in half. During the summer months, the elite decamped to the countryside where, historians generally suggest, they dedicated themselves to rural pursuits, particularly riding and hunting.[7] Yet, on returning to town for the winter season, we have been led to believe, they relinquished these diversions and devoted themselves to sociability, conducted for the most part indoors and involving a frantic round of balls, plays and concerts.[8] Some scholars have argued that rural life, and particularly its sports, appealed to men, in conflict with a female preference for urban diversion. Others have emphasised that the provincial season afforded both sexes considerably more than sport.[9] Yet, despite these revisions, a wealth of evidence showing that riding played a major role in the London season has been overlooked. This may stem from eighteenth-century stereotypes which polarised town and country. By the 1740s, when Henry Fielding lampooned hunting-mad squires in *Joseph Andrews* and *Tom Jones*, recreation was on the front line of the culture clash between town and country.[10] And throughout the second half of the century, writers continued to juxtapose the rustic bumpkin, forever muddy and going off to hunt, with the well-dressed Londoner ensconced indoors. In one camp, urbane intellectuals led a vitriolic attack on rural field sports; in the other, provincial writers such as William Cowper claimed that London's routs and masquerades bred affectation and deceit, while the city's gaming tables and bagnios polluted the nation's morals and bankrupted its leaders. Cowper, however, retained his faith in Londoners, proclaiming 'That man, immured in cities, still retains / His inborn inextinguishable thirst… / To range the fields, and treat their lungs with air'.[11] And as will become clear, one of the most effective and appealing ways that Londoners found to do this was on horseback, or in a horse-drawn vehicle.

Another striking feature of the historiography surrounding London's recreational life is the extent to which sociability has been foregrounded. This is problematic because while the city's elite habitually judged venues and events on the quality of the company, and the conversation which flowed from it, by focusing so heavily on sociability, historians have neglected other motivations which may have transcended class divisions. Today, it is widely accepted that animals can become highly valued companions. People engaged in such relationships are thought to benefit from the opportunity to care for something, the recognition of another being, a sense of security and stress relief.[12] Historians have started to explore these dynamics in pockets of time and place, with most attention being given to pet-keeping, but both Georgian London and horse riding demand much greater attention.[13] This chapter begins by showing that equestrian culture became one of London's most successful recreational departments in the second half of the eighteenth century. It then examines the role played by horses as facilitators of sociability before considering the possibility that these animals provided an alluring alternative to sociability.

As shown in the previous chapter, London's wealthy elite stabled increasing numbers of riding and carriage horses in West End mews in the eighteenth century. This facilitated rising participation in equestrian recreation but in the second half of the century, ownership of an equipage also became less of a barrier, as stable-keepers and horse dealers began to hire out horses and vehicles at more accessible prices. The daily rate for a saddle horse in the 1780s was 6 or 7s while a one-horse chaise cost 10s 6d.[14] This was expensive compared to the 3s charged by the Drury Lane theatre for a place in the pit but such services were attractive to lesser gentry and upper middling sorts who either could not afford a private equipage or required additional horses and vehicles on a short-term basis. Moreover, much lower down the social ladder, many men who were unable to rent found other ways to ride. Some male servants were allowed to exercise horses belonging to their masters, while tradesmen and labourers occasionally commandeered cart horses and other working animals. The West End elite continued to enjoy privileged access to certain facilities and activities but the dramatic expansion of London's horse stock and the commercialisation of equestrian recreation gave many more people opportunities to ride. These developments were felt by women as well as men, although equestrian culture remained deeply patriarchal. Wealthy men held sway over the family equipage, controlling both its financial affairs and use; aristocratic men governed turf and field as investors and rule-makers; jockeys were always male;

by the 1750s, the mounted huntress was an endangered species; and the achievements of a riding fraternity were almost exclusively written about by men.[15] Nevertheless, as we will see, many polite and middling women seized opportunities to participate in some of London's equestrian activities.

Commercialisation produced numerous equestrian venues in the metropolis, each with its own distinctive character, clientele and riding behaviour but we should begin with the nerve centre of elite equestrianism: Hyde Park. After the Restoration, riding was central to the park's renaissance as a place of public resort. Its focal point was a 'ring', about 250 paces across, which was railed off for the use of carriages, and soon proved a magnet for the West End's fashionable new residents. In 1711, *The Spectator* complained that a mob of servants was 'let loose' at the entrance to the park 'while the Gentry are at the Ring'; and in 1732, Thomas Salmon observed that with the arrival of summer 'we frequently see four or five Lines of Noblemens and Gentlemens Coaches, rolling gently round the Ring in all their gayest Equipage'. By then, however, it would appear that some riders were growing tired of 'aimless circuits' – in 1736, one writer found the ring 'quite disused by the quality and gentry' – and soon after the site was partly destroyed to create the Serpentine River.[16] In all other respects, the royal scheme dramatically increased the park's equestrian appeal: a new *route du roi* was dug to replace that constructed by William III and rather than turn the King's Old Road back to turf, it was retained as a public riding parade. The *beau monde* quickly embraced what, by the early 1760s, had become known as Rotten Row.[17]

By the 1790s, public riding had spread to the King's New Road, more commonly referred to as South Carriage Drive or simply the Drive. In April 1791, the *Argus* reported that

> Exercise in *every way*, it seems, is the thing now on *Sundays*, amongst the bucks of fashion. The true ton is, to *grind gravel* with your curricle in the *drive*, from half past two till half past three; then stretch your leather in the *ride* [Rotten Row], till half past four.[18]

The park now swarmed with riders, vehicles and gawping pedestrians. One first-time visitor from Lancashire observed: 'The number of Carriages is truly astonishing; for the whole length of Hyde Park which, in one view, I conceive cannot be less than a Mile from three to five o'clock you may see Carriages two fold continually passing.'[19] This is the equestrian paradise captured in *The Entrance to Hyde Park on a Sunday*, an engraving published in 1804 in the popular guidebook, *Modern London* (see Figure 28). Riders pack Rotten Row as far as hundreds of admiring spectators can see. In the foreground, fashionable carriages pour in and in the distance, unsaddled horses graze and frolic. The explosion of equestrian activity witnessed in this period was the catalyst for a grand scheme to equestrianise the park – by 1827, Rotten Row and the Drive had been integrated into a complete formal circuit including an Outer Ring, East, North and West Carriage Drives and the Serpentine Road.[20] Despite being free to enter, Hyde Park played a key role in the commercialisation of equestrian culture because it was here that the most

28 John Pass after Edward Pugh, *The Entrance to Hyde Park on a Sunday*, engraving published in Richard Phillips, *Modern London; Being the History and Present State of the British Metropolis* (1804).

fashionable riders came to show off their horses, vehicles, tackle and riding apparel. At the same time, the park was the principal stage on which to demonstrate equestrian ability and from the mid-eighteenth century this was increasingly acquired in London's public riding schools.

These institutions were considerably less decorous than many other polite venues but they were significant players in the commercialisation of leisure. In the 1750s, elite equestrians, including the tenth Earl of Pembroke, gave their patronage to continental riding masters, among whom the Frenchmen Henry Foubert, his nephew Solomon Durrell, and an Italian, Domenico Angelo, opened the city's first public riding schools. These pioneers promoted a continental system of managed riding, or what we might loosely describe today as dressage. They were encouraged by George III, who constructed a private riding house at Buckingham House in 1763–66 and inspired the Dukes of Cumberland and Gloucester to follow suit. In 1771, Richard Berenger, the king's gentleman of the horse, published a guide commending the manège to modern riders but he was too late, they had already moved decisively towards the 'English Hunting Seat', a system that promoted ease and the riding of increasingly fast horses over fences in the English countryside.[21] London's riding schools were not just swift to adapt to this transition, they took a leading role in promoting it. Newspaper adverts reveal that between 1760 and 1835, the city produced twenty-six new public riding schools. On average, these businesses advertised for twenty-seven years although the most successful did so for more than forty years. The number of schools increased without interruption from two in 1760 to thirteen in the late 1780s. There was a downturn during the Napoleonic Wars when a sharp rise in the cost of provender and a contraction in consumer spending forced all but six establishments out of business, but the trade rallied in the 1820s, returning to a total of twelve. As this golden age drew to a close, the number of schools dipped to nine in the early 1830s but resisted serious decline until the 1850s.

The geographical distribution of London's riding schools over time is revealing. The years 1761–65 saw an initial expansion which broke the monopoly of Foubert's Royal Academy at Golden Square, Soho. Two new schools opened nearby while a third was established in Moorfields to meet growing mercantile demand. In the years 1786–90, the number of schools almost trebled, with Hyde Park Corner becoming the principal hub. These new schools sought patronage from the wealthy new residents of Mayfair and Knightsbridge while also benefiting from the growing popularity of park riding. At the same time, new commercial wealth fuelled demand for schools in Moorfields, Whitechapel and Lambeth. Each of these sites developed an array of services in the second half of the eighteenth century. Proprietors developed pricing structures that protected their respectability by excluding most tradespeople while welcoming middling clients. In the 1780s, Carter's near Grosvenor Square charged £2 7s for twelve lessons taken at a pupil's convenience, while a decade later, the Pantheon Riding School on Blackfriars Road charged £3 13s 6d for an intensive course of twenty-four sessions as well as the more affordable option of single lessons priced at 5–7s. Some houses levied an annual entrance fee: Angelo's Academy in Soho charged a steep three guineas in the 1760s but one guinea was sufficient to join Park & Son's less exclusive school in Upper Moorfields in the 1780s. Other establishments offered free entry to ride within certain hours, a strategy that promoted more lucrative services including lessons, stabling, breaking-in and selling horses on commission.[22] Carter's probably developed the most comprehensive subscription system: six guineas secured riding instruction three times a week for eight months, the breaking-in and exercise of one horse and use of the riding house. Establishments with enough horses also hired them out for short periods.[23]

Like many other recreational venues, London's new riding schools made astute use of newspaper advertising to promote their services and were alert to emerging equestrian trends. By 1783, Charles Carter was already teaching ladies 'to rise in the stirrup' to ease jolting when riding on the road, while Philip Astley advertised that he broke horses 'for the Army, Road, Field and Draft; also for Stalking, Shooting, and particularly for Ladies Riding'. When magistrates forced Astley to defend the legitimacy of the circus in the 1770s, the fact that his performers also taught respectable riders, broke-in horses and trained them for the battlefield reassured the elite that the exuberant drama had rational applications.[24] A major achievement of London's riding schools was to assimilate the discipline of managed riding with the ease of jockey riding to suit the needs of individual riders, male and female, child and adult, sporting and sedate, rural and urban. Rather than using the language of the manège, proprietors developed a new vocabulary for a modern polite art of riding. This prioritised three outcomes: 'a graceful appearance on horseback', the demonstration of 'ease', and the prevention of accidents.[25] In promoting deportment, riding masters shared something in common with metropolitan dancing masters and instructors in swordsmanship. Their skills continued to be valued in polite society throughout the eighteenth century, as shown by a letter sent by the fourth Earl of Chesterfield to his son at Westminster School in 1751. The earl implored his nineteen-year-old son not to 'neglect your exercises of riding, fencing, and dancing... for they all concur to *dégourdir* [smooth rough edges], and to give a certain air. To ride well, is... a proper and graceful accomplishment for a gentleman'. This could have been an advert for Angelo's Academy, which offered intensive lessons in all three arts in the 1760s and 1770s.[26] Riding schools sought to maintain patronage by making architectural improvements and taking greater efforts to ensure rider safety. In the late eighteenth century, the Pantheon school boasted that it was 'the most warm and commodious of any in or near the metropolis' while Carter reassured his clients that stoves kept a 'place for ladies attendants' warm even in damp or cold weather. Many proprietors banned dogs to avoid the kind of disruption sketched by Thomas Rowlandson in the late eighteenth century (see Figure 29), and promised to mount nervous riders on animals 'purposely broken in to accommodate every capacity'.[27]

The commercialisation of equestrian education produced some highly successful businesses at the same time as horse repositories such as Tattersall's were making their mark in the metropolis. In the 1760s, the *bon ton* was particularly well represented at Hall's on Hyde Park Corner, with clients including Lady Mary Coke, the Duke and Duchess of Bolton, and the Duke of Roxburghe. In 1787, Carter offered prospective clients a printed list to prove that 'Nine hundred ladies of the

29 Thomas Rowlandson, *The Riding School* (pen and brown ink over graphite, undated).

highest rank and fashion, and nearly an equal number of Gentlemen have been Taught to Ride'. His proprietorship began in 1778, suggesting that the school attracted almost 200 students a year. By then, Angelo was at the height of his fame and was thought to be earning £4,000 a year. In the early 1800s, Fozzard's of Park Lane became the most fashionable school in the city and eventually secured the patronage of a young Princess Victoria in 1831.[28] Riding schools were influential institutions in polite metropolitan culture, made more so because they forged such close links with other strands of equestrian recreation. This included teaching Londoners how to ride in the countryside.

'Riding out' could be enjoyed on horseback or seated in a light vehicle and involved five interrelated activities: airings, exercise, commuting, race-going and hunting. Londoners had been riding in their suburbs for centuries but the practice became increasingly prevalent, multifaceted and commercialised after 1750. Historians often associate the rise of domestic tourism in the mid-eighteenth century with the inaccessibility of the Continent at this time but it was also fuelled by a passion for riding which steadily intensified over the course of the Georgian period.[29] In common with other equestrian activities, riding out encouraged spending on horses, vehicles and tackle, but avoided the fees charged by riding schools and the costs associated with extravagant display in Hyde Park. As a result, riding out was particularly attractive to London's middling sorts. Previous studies of middling and particularly mercantile lifestyles have tended to foreground venues such as coffee houses, livery halls and voluntary societies, but the Square Mile produced some of London's most enthusiastic riders.[30] The mid-eighteenth-century diaries of John Eliot and Thomas Bridge provide a window onto this world and suggest that horses were central to the recreational lives of at least some middling men. In 1757, Eliot was an unmarried, twenty-two-year-old insurance underwriter living in Bartholomew Close, Smithfield. He owned his own horse, which he kept at livery in Coleman Street, Lothbury, a convenient ten-minute walk from his house. Despite his heavy workload, Eliot usually rode once a week and found opportunities to do so on every day of the week except Sunday, when he attended Quaker meetings. Half of his rides took place on Saturdays and a quarter on Thursdays. He set out on half of his rides mid-morning, after reading and taking breakfast, but also mounted mid-afternoon on a third of occasions. In 1762, Thomas Bridge was a twenty-three-year-old drug merchant living above his counting house in Bread Street, near Cheapside. Unmarried and in the early stages of building a substantial fortune, Bridge dedicated his recreational life to riding and visiting suburban friends. He rode out more than once a week, mostly at the weekend, usually after dinner but also after breakfast. In the first quarter of 1762, Bridge spent over £100 on a chariot, two brown geldings and a chaise with harness and trunks. He maintained this impressive equipage at The George, a large coaching inn on Snow Hill, Holborn.[31]

Bridge and Eliot's behaviour suggests that mercantile citizens developed a distinctive riding culture which although influenced by aristocratic example remained independent of it. At a time when Hyde Park riding was becoming increasingly fashionable, Eliot never mentioned riding there while Bridge did so only once. Having ridden to the West End to examine a chaise, Bridge found that the vendor, the Bishop of

Bristol, was still in bed and was forced to while away a couple of hours in the park. While some polite commentators feared that merchants and tradesmen were invading the park, there is scant evidence for this. Park riding retained an aura of West End exclusivity throughout the period, while wealthy citizens continued to express a strong preference for riding out. This is hardly surprising considering that getting to Hyde Park from the City involved a 5 km ride through some of the capital's busiest streets and even the short distance from Soho Square was enough to make one polite youth complain that 'Riding over stones is very disagreeable'. The *beau monde* may also have made self-made visitors feel unwelcome but there were more positive reasons for citizens to ride out. Analysis of this culture underscores Perry Gauci's assertion that mercantile actions should not be interpreted 'from the perspective of a landed society, without recognition of more immediate foci for City loyalty'.[32]

Riding played an important role in consolidating the group identity of affluent middling sorts, particularly its men. By the late eighteenth century, the sight of 'cits' riding in the suburbs had come to symbolise their growing confidence and inspired one of the best-known caricatures of the age, John Gilpin. First published in 1782, William Cowper's comic ballad, *The Diverting History of John Gilpin*, follows a Cheapside draper's ludicrous cross-country dash on a runaway horse to his villa in Hertfordshire.[33] Riding out gave Cowper the ideal device to mock urban life and its invasion of the countryside but metropolitan satirists were more concerned about the social status of those taking part. In *Cit's Airing Themselves on a Sunday*, published in 1810, for instance, Thomas Rowlandson depicts a riding party led by merchants mounted on a pair of unattractively squat hacks (see Figure 30). The riders are well-dressed but the burliness they share with their horses immediately gives them away. The image served as a critique of social emulation, appealing to elite snobbery towards an increasingly self-assured mercantile class. In a similar vein, one writer derisively claimed that

> a swarm of young clerks in office, and in banking and commercial houses, the moment business is over, issue from the city, cloathed in excellent imitation of men of fashion; some booted and spurred as if they had been

30 Thomas Rowlandson after Henry William Bunbury, *Cit's Airing Themselves on a Sunday* (hand-coloured etching, 1810).

> riding all the morning, parade up and down… [Bond Street] and jostle bucks of rank and fortune.[34]

In reality, horse riding enthralled many middling men and played a central role in their professional, recreational and family lives. As early as the 1720s, Defoe could marvel at the elegant suburban villas springing up in Essex and Surrey, from which increasing numbers of merchants and financiers commuted into the City. While some chose to walk or took advantage of an expanding network of short-stagecoaches, many rode on horseback or travelled in their own chaise. Thomas Bridge's emergence as a commuter in the 1770s marked the end of his youth. Having married and become a father, he acquired a villa in Tottenham and substituted his free-ranging jaunts for a far more repetitive commute supplemented by occasional Sunday outings to villages near Tottenham. The need to obtain news and make contacts meant that merchants were under constant pressure to keep on the move. This made riding an ideal recreational choice for these men, providing they could maintain a healthy balance between work and pleasure. In the 1750s, John Eliot often managed to use predominantly recreational rides to complete small acts of business, carefully planning his route so that he could collect or deliver paperwork as well as profit from exercise, scenery and fresh air.[35]

Riding out was closely linked to two other popular middling activities: race-going and hunting. In the second half of the eighteenth century, thousands of metropolitan riders flocked to racecourses in the south-east every year. Between the 1680s and the 1730s, horse racing expanded on an unprecedented scale across England, not least around the capital. In 1700, Barnet, Croydon and Epsom had been the only racecourses within 32 km of the city but by 1738, there were no fewer than nineteen. Those at Tothill Fields, Kentish Town, Belsize, Hampstead and Highgate were 1–8 km from Charing Cross. Ten courses were 8–24 km away, including Finchley, Wimbledon, Barnet, Enfield, Hounslow, Croydon and Epsom. Four more, Epping in Essex, and Cobham, Limpsfield and Egham, in Surrey, were 24–32 km away. This spawning came to a sudden halt in June 1740 with the introduction of an Act 'to restrain and prevent the excessive increase of horse races', which stipulated that all prizes had to be worth at least £50 or more. Across the country, the legislation culled the vast majority of races. London's wealth softened the blow but only six local venues survived after 1740.[36] After a decade in recession, however, English racing began to recover and commercialisation gained momentum. In the metropolitan area, the number of races held at the surviving courses increased dramatically, as did the prize money. The crisis eliminated the more parochial meetings, leaving behind the bigger players so that by 1760, there were only three sites within 32 km of the metropolis: Barnet, Epsom and Egham. An important consequence of these developments was a major increase in the number of Londoners riding out to races. In 1771, one newspaper estimated that more than 30,000 people had attended Barnet races in a single week, the majority Londoners.[37] Before 1740, most courses had been within easy walking distance of the city but by the 1760s, the nearest, Barnet, was 16 km away, while Epsom and Egham were 22.5 km and 30.5 km away respectively. Henceforth, as a rip-roaring song from the 1760s makes clear, riding out became the primary mode of going to the races. Beseeching all Londoners to saddle up, 'Invitation to Epsom Races' extolled the complementary pleasures of riding out and race-going. These included sociability and display, escape from urban life, health benefits and the thrill of speed:

> Come Nobles, and Heroes, and Bucks of the Turf;
> Having had of the dull smoaky Town quite enough;
> Come mount the gay Steed; and to Epsom repair,
> To see the fine Horses, and Ladies, so fair!
> Come Statesmen so subtle, unbend for a while,
> And leave your deep Schemes, on our Races to smile,
> In your Coaches, so splendid, at Races preside,
> And learn of our Jockeys how People to guide.
> Come Merchants, and Bankers, and Poets, and Players,
> Leave your discounting Bills, and your anxious Affairs
> Come mount the proud Steed, and to the Races advance,
> To taste Health and Pleasure, not equall'd in France...[38]

During the racing season, London's stables and repositories did a roaring trade in horse and vehicle rentals. In 1785, the *Whitehall Evening Post* reported that Epsom was 'visited by phaetons, curricles, tim-whiskies, gigs, buggies, and sulkies, out of number' and, less plausibly, 'pleasure-carts drawn by Jack-asses'. And in 1792, the *Evening Mail* observed that 'Epsom Races were never more full... Not a chaise was to be had yesterday for love or money. The road was lined all the way [from London] with carriages'.[39] The commercialisation of racing created a new impetus to ride out and strengthened the bond between London's equestrian and sporting departments, the most culturally significant demonstration of which was hunting on horseback.

In 1826, a practical treatise on fox hunting paid special tribute to London and its riders. The book's author, a seasoned veteran of hunts in Essex and Suffolk, assured those who 'keep hounds at no great distance from London' that they would

> find many of the inhabitants of that capital (cockneys, if you please) *good sportsmen*, well mounted, and riding well to hounds: they never interfere with the management of them in the field, contribute liberally to the expense, and pay their subscriptions regularly.[40]

This well-meaning, if somewhat patronising, defence of metropolitan sportsmen acknowledged decades of rural snobbery. As early as 1707, a ballad sardonically described the City's Easter Monday hunt in Essex as 'a most pretty show', after which the riders returned to London with 'Their faces all so torn and scratch'd, their wives scarce knew them well'.[41] By the early 1800s, the sporting Londoner had become a familiar caricature and in the 1830s John Jorrocks, 'a cit rapturously fond' of fox hunting, emerged as one of Britain's most popular comic creations.[42] Robert Surtees' *Jorrocks's Jaunts and Jollities* (1838) revelled in the 'eccentric and extravagant exploits' of a 'substantial grocer' from St Botolph's Lane, and his fictional rides with the real Surrey Hunt. These episodes were plagued with calamity, humiliation and suspect achievements. On one occasion, Jorrocks plunges into a cesspool, on another he crosses the Croydon canal on a barge after his horse refuses to jump. Surtees lampooned cockney sportsmen for talking shop, lacking equestrian ability and quitting the field early, but he was merely teasing his younger self and the friends who saved him from the drudgery of legal work in the 1820s. In doing so, Surtees satirised rural snobbery and celebrated the end of a golden age of metropolitan hunting, the culmination of seven decades of prosperity.[43] By the early nineteenth century, London was producing heroes of the chase to rival those of Leicestershire and Yorkshire, including the eccentric Colonel Hylton Jolliffe MP, master of the Merstham foxhounds in Surrey. When Jolliffe's pack was disbanded in 1830, the *Sporting Magazine* reflected:

> Who has not seen him walk up St. James's Street with… [his] neat blue coat with metal buttons… and clean yellow leather shorts with long gaiters? He looks like what he is, a country gentleman and a fox-hunter.[44]

The capital's wealth underpinned the success of fox hunting in the south-east because the infrastructure needed to hunt on horseback was so expensive. The 1826 treatise commended the city's fine horses and commitment to retaining local packs of hounds funded by kennel subscriptions. Most of these had emerged in the final quarter of the eighteenth century and proceeded to jostle for territory and riders drawn from the capital's resident nobility, gentry and middling sorts.[45] Surrey first became a hub of metropolitan fox hunting in the 1760s, when a pack of hounds was kept at Bermondsey. They were soon followed by hounds belonging to Mr. Walker of Putney, which pursued foxes in Wimbledon, and the establishment of the Surrey Hounds at Godstone.[46] A second pack, kennelled at Leatherhead, hunted further to the west and at the end of the century, they merged to form the Surrey Union. By then, Londoners enjoyed an overwhelming choice of suburban hunts. In 1796, the *Sun* newspaper predicted that

> our sporting friends, in and near the Metropolis, will have no reason to repine at the present prospect of sport for the season, as Wood's harriers take Sunbury [and] Hounslow Heath… whilst on the other side of the Thames, Kingston Hill, Wimbledon Common &c. is possessed by the excellent pack of Mr. Chapman. The adjoining Country is covered by Mr. Gee, of Beddington… to these are added… the King's stag hounds and harriers… [and] Lord Derby's at the Oaks [near Banstead].[47]

One of those to enjoy these opportunities was William Cobbett's adolescent son Richard. Cobbett wrote in 1825 that he 'can ride… over anything' and 'begins to talk of nothing but fox-hunting!' Such enthusiasm was worthy of the great hunting families of the north and Midlands but Cobbett senior ran a plant nursery in suburban Kensington and his son pursued foxes within a few kilometres of the metropolis.[48]

The commercialisation of metropolitan fox and stag hunting both encouraged and benefited from developments in horse dealing and racing, as well as the advent of riding schools offering lessons and breaking-in specifically for the field. Nevertheless, the city's sportsmen faced a growing problem. In the 1770s, hunts often took place within 8 km of the Thames, but over the next half-century urbanisation made this increasingly untenable, as shown by occasional newspaper reports of farcical incidents. In 1788, for instance, the *World* observed that the Surrey Fox Hounds had been forced to dig out a fox after it ran to ground 'under the Tower on Shooter's Hill', a popular suburban resort. In a similar vein, in one episode of *Jorrocks's Jaunts*, the Surrey Stag Hounds catch their quarry 'once in a mill-pond, once in a barn, and once in a brick field'. Surtees joked that this was all part of the experience but by the early nineteenth century, most metropolitan hunts had moved deeper into the countryside, with chases often finishing 30 km from the city. This required further investment and organisation because the need to ride out several kilometres simply to join a meet threatened to exhaust horses before the chase had even begun and undermine the pleasure of a day's sport.[49] Some Londoners began to hire fresh horses from suburban 'hunting stables' like the Derby Arms at Croydon, where they left their own animals to stand at livery. But as hunts strayed further from the metropolis, sportsmen discovered the perils of returning home in the dark. In 1782, a Park Lane stationer was robbed by a highwayman after hunting around Beaconsfield in Buckinghamshire, 32 km from town. By the time he reached Bayswater, it was already dark.[50] A novel solution for those who could afford it was to hire or even purchase a compact residence in 'good sporting country'. Properties advertised in the London press in the 1790s included a 'Sporting Box' complete with a coach house and stabling for nine horses on the borders of Epping Forest. At the 'very easy and convenient distance' of only 22 km from London, the premises also benefited from packs of stag- and fox-hounds being kept in the vicinity.[51] The sporting box epitomises the wealth and ambition of London's equestrian culture in the late eighteenth century, but we still need to unpack what made Londoners so passionate about riding horses.

Sociability was a driving force in polite metropolitan living and horses served this culture by maximising the number of visits, balls, plays, routs and dinners that an individual could attend in the season. But as I have suggested, horse riding was a sociable activity in its own right. Riding schools were not only temples to equestrian pleasure and perfection, they were refined resorts. Proprietors worked hard to create an ambiance conducive to polite conversation, crucial to which was a comfortable viewing gallery. In the 1760s, Lady Mary Coke timed her visits to Hall's riding house specifically to meet friends and acquaintances; and in 1831, the young actress Fanny Kemble enjoyed 'a pleasant, gossiping ride with Lady Grey and Miss Cavendish' at Fozzard's. On another occasion, these women discussed the theatre and the 'stay-at-home sensation', which they condemned as an unsociable fad.[52] Riding schools did impose certain limitations on sociability, particularly between men and women. The intense physicality of riding, its potential to ruffle garments and expose flesh, constituted a potential threat to decency and decorum, particularly when so many pupils were unmarried girls, meaning that proprietors had to draw up strict rules and timetables to exclude voyeurs.[53] Nevertheless, these venues successfully promoted sociability among elite and upper middling riders; and as pupils graduated to other equestrian diversions, these opportunities multiplied.

Nearly three-quarters of the rides which the underwriter John Eliot took out of the city incorporated some kind of sociable recreation such as tea drinking, dining at an inn or visiting friends and relatives; and almost all of Thomas Bridge's outings in 1762 led to drinking at taverns, playing bowls, sightseeing or strolling with friends. Rowlandson satirised this behaviour in *Cit's Airing Themselves on a Sunday* (see Figure 30), which features a couple flirting on horseback while an elderly Jew, his wife and their daughter are crammed into a chaise. A middle-aged volunteer cavalryman brings up the rear. Riding out certainly promoted mixed-sex sociability but this image cannot be read as typical. Of Bridge's outings, 40 per cent led to some form of interaction with women but when he actually rode in company it was almost always with the same man, his friend Samuel Kirkman. Providing they were physically able, men usually opted to ride out on horseback rather than in a vehicle – Bridge owned a chariot but made two-thirds of his outings in the saddle – but women almost always travelled in a chaise.[54] This promoted physical segregation but also drove an experiential wedge between the sexes. There were some exceptions – in the 1760s and 1770s, Lady Mary Coke regularly rode out alone and on horseback from her house in suburban Notting Hill – but by the 1750s, virtually all female riders in Britain rode side-saddle, which made long-distance travel uncomfortable and precarious. These obstacles diminished riding out's credentials as a mode of sociability. The most successful equestrian activity in this regard was park riding.[55]

Hyde Park was a place to be seen and the best way to achieve this was by riding on horseback or in a horse-drawn vehicle. Sylas Neville, a twenty-eight-year-old bachelor of modest fortune, was acutely aware of this when, in May 1769, he decided to ride his first horse 'round Hyde Park in his new bit made from Lord Pembroke's pattern and Hussar saddle'. Eager to be acknowledged and accepted by the *beau monde*, Neville hoped to exhibit refined taste. Having spent months selecting his horse, Neville adorned it with tackle inspired by elite equestrian example. At the same time, he hoped to demonstrate equestrian prowess, to which end he had enrolled with a West End riding school.[56] In theory, elite equestrians shared the same rides as their social inferiors because the park was open gratis to anyone 'of decent habit and demeanour', but as Hannah Greig has cautioned, social exclusivity in London was often deftly conducted within seemingly inclusive public arenas.[57] This was certainly the case in Hyde Park where equestrians rode in distinctive ways and mounted different types of horse to project their status. Lady Mary Coke offers a revealing snapshot of this behaviour in a diary entry from May 1767, which records that she

> came into Hyde Park [and] rode all the time with Lord Bathurst... Just as we came home we met the Duchess of Norfolk. She stopped her Chaise, & desired me to dine with her. I accepted the invitation, & came home.

Riding was an effective way to maintain exclusivity. Elite equestrians often entered Hyde Park together or orchestrated desirable encounters by looking out for familiar horses, vehicles or liveries, and riding to intercept a favoured companion. By drawing up, the duchess acknowledged Coke's acceptability and initiated conversation. By contrast, undesirables could be excluded by riding off or abruptly switching direction.[58] At the same time, riding styles, as well as the appearance and behaviour of horses, were just as useful for discriminating between different ranks as an individual's clothes, manners and conversation. In *Sunday Equestrians or Hyde Park Candidates for Admiration*, 1797, Isaac Cruikshank identified eight satirical case studies (see Figure 31). At one end of the spectrum, he depicted a well-dressed gentleman mounted on a fine white stallion, apparently trained in the manège, and a lord with an impressive skewbald thoroughbred. At the other, he mounted a bumpkin on a shaggy country horse and forced a hapless gentleman to ride a dray horse

31 Isaac Cruikshank, *Sunday Equestrians or Hyde Park Candidates for Admiration* (hand-coloured etching, 1797).

because his thoroughbred mare was lame. And at the very bottom of the pile, Cruikshank presents a lean tradesman riding a jackass which he has 'crop't and docked' in hopes 'they'll admit him into the park'.

For park riding to contribute to sociability, conversation had to flow freely but managing horses, whether from the saddle or in a vehicle, required concentration and even well-trained animals could interrupt by disobeying, shying or bolting. In 1763, the Prince of Wales narrowly escaped a serious accident in the park when one of his horses 'took fright and got his leg over the pole of the coach; which set the horses a plunging'.[59] The sociability of a recreation was also judged on the quality of conversation and obsessive talk about riding or hunting was associated with dull provincials. Adept park riders had to overcome these challenges as Lady Mary Coke and Lord Bathurst did in 1767: with Coke in a chaise and Bathurst in the saddle, the pair politely conversed about 'some of his old acquaintances, Mr Pope, Swift, [and]

Lord Bolingbroke'.[60] There was nothing scandalous about this interaction but mixed-sex sociability did provoke considerable anxiety at times of national insecurity and park riding, in particular, was seized upon as an example of London's frivolous and effeminate ways.[61]

Georgian England held accomplished male riders in high esteem but equestrian ability was judged on fluid and contestable criteria. In the second half of the eighteenth century, some commentators viewed park riding as a threat to more useful and manly strands of equestrian culture such as field sports.[62] The most contemptible expression of this was the macaroni rider, a male figure who mounted simply to parade his flamboyant dress and a Frenchified mode of riding. In the mid-1790s, when the macaroni had lost most of its cultural resonance, John Lawrence noted that some continued to ride

> up Rotten Row bolt-upright… as though he were impaled, his stirrup-leathers of an excessive length, the extremity of his toe barely touching the stirrups… his lily hands adorned with ruffles Volant, and his head with a three-cocked hat.[63]

Taking part in polite society often challenged traditional male values but this was particularly evident when it came to riding because elite Englishmen had traditionally expressed so much of their virility and authority through horsemanship. It was in this context that James Boswell chided Dr Johnson for being 'a delicate Londoner', adding 'you are a macaroni; you cannot ride'. This was in 1773 and the friends had begun a tour of the Highlands. As they left Montrose, Boswell worried that the sixty-four-year-old was flagging but Johnson retorted 'Sir, I shall ride better than you'. Boswell's journal of the tour was first published in 1785, by which time the macaroni rider – a creature of Hyde Park lacking the skill and strength to ride over real countryside – was firmly established in British culture. These tensions smouldered in the midst of a national identity crisis triggered by the American Revolutionary War.[64] In 1779, Carington Bowles published *Kitty Coaxer Driving Lord Dupe Towards Rotten Row*, a satirical mezzotint inspired by the life of the courtesan Kitty Fisher, who was thrown from a horse in 1759 as she cantered through St James's Park (see Figure 32). Unhurt, Kitty was

32 Anon., *Kitty Coaxer Driving Lord Dupe Towards Rotten Row. From the Original Picture by John Collet, in the Possession of Carington Bowles* (coloured mezzotint, 1779).

said to have burst into a fit of laughter until she was picked up by a fine chaise and swept through the crowd of onlookers. By inverting accepted gender relations, this image urged men to seize the reins and defend their country.[65]

Mixed-sex sociability in the park came under renewed scrutiny in the wake of the French Revolution. Between 1790 and 1796, several poems, songs and novels depicted dissolute male riders in hot pursuit of women.[66] Most were gentle in their disapproval but attitudes appear to have gradually hardened during the French Revolutionary Wars. Between 1794 and 1804, a series of invasion threats convinced hundreds of wealthy Londoners to fund, train with and lend horses to volunteer cavalry regiments in the metropolis. Riding in such regiments involved relatively little danger but to be seen doing so was a public affirmation of an individual's masculinity and patriotism.[67] Large crowds came to see and cheer on training exercises and reviews in Hyde Park, St George's Fields and Wimbledon Common.[68] Seen against this backdrop, bucks chasing after women in Hyde Park appeared to some a shameful betrayal of patriotic duty. Addressing Parliament in 1798, Sheridan railed against 'our young men of fashion' who 'might be better employed in contributing to the defence of the country, than in… taking the field in Rotten-row'. As Linda Colley has shown, this was a period in which aristocrats made various efforts to refashion their image from 'parasitic' to patriotic servants of the state; and being one of the most visible expressions of elite culture, equestrianism came under particular pressure.[69]

Advanced furthest by park riding, sociability was facilitated by every strand of London's equestrian culture. But this does not explain why many Londoners appeared to prefer riding to other sociable activities. Georgian commentators often complained that assemblies, dinners and routs were spoiled by dull conversation and stifling company so we need to consider the possibility that interactions with horses offered something which people did not.[70] We can glean a sense of this from the correspondence of a twenty-year-old Fanny Kemble, a member of London's leading theatrical family. In January 1830, she wrote excitedly to a female relative:

> my dearest H… I am exceedingly happy… my father has given me leave to have riding lessons, so that I shall be in right earnest 'an angel on horseback,' and when I come to Ardgillan… I shall make you mount upon a horse and gallop over the sand with me; won't you, my dear?[71]

With the London season in full swing, it is remarkable that Kemble expressed more excitement about riding than she did about theatregoing, concerts or balls. She enjoyed 'gossiping' with other young women when she was riding at Fozzard's school but this was not her principal motivation. Writing in her diary two years later, she described riding as an exhilarating recreation in its own right. By then, Kemble had become a spirited rider determined to test her nerve and skills. During a visit to Hyde Park in January 1832, she performed her sociable duties by walking 'soberly round the park' and speaking to 'friends and acquaintances' but then mounted a 'great awkward brute' of a horse and 'determined once more to try… [his] disposition… I flitted down Rotten Row like Faust on the demon horse'.[72] For Kemble and many other polite Londoners, horse riding kept dullness at bay but more than this, equine companionship could become an alluring alternative to sociability. This becomes particularly clear when we encounter Londoners riding out of town, behaviour which signalled a desire to escape the city but also to test equestrian ability.

Many of the locations favoured by metropolitan riders were established resorts such as Hampstead, Muswell Hill, Dulwich and Putney, to which were added several picturesque villages including Hackney, Tottenham, Edmonton, Romford, Battersea and Clapham. Even in the early nineteenth century, rides to these spots would have been dominated by traversing fields, heath, marsh and woodland. Some rode out 16 km or more into the countryside before turning back, while others spent a night or two in a rural inn or at a friend's house.[73] John Eliot rode to seventeen destinations scattered 5–23 km from the City. Surrey featured most often, partly because his grandfather lived in Croydon, but he also favoured the scenic villages of Clapham and Putney. A quarter of Eliot's rides took him to villages north of the City, especially Tottenham and Enfield. As a bachelor, Thomas Bridge was even more adventurous. He displayed a strong preference for riding through the Hackney and Leyton Marshes but reached thirty-three destinations up to 35 km from the City. Horses liberated Londoners by expediting access to private space but also because they greatly enhanced the experience of exploring the countryside. By elevating riders, horses offered privileged access to the kind of 'unbounded views' which Joseph Addison deemed so important because they gave the eye 'room to range abroad… and to lose itself amidst the variety of objects that offer themselves to its observation'. Riders could transform their perspective at whatever speed they chose and if still not satisfied, they could instruct their horse to jump obstacles for added exhilaration.[74]

It is easy to imagine why this would have been so appealing to individuals who spent much of their lives in the City's claustrophobic streets. Riding out substituted pollution for good country air, as well as stimulating circulation and loosening muscles, which were widely accepted remedies for relieving stress and lethargy. Some of the most popular medical guides of the eighteenth century asserted that riding prevented and alleviated ill-health. In the 1720s, the physician George Cheyne proclaimed that 'The Digestion and the Nerves are strengthened, and most Head-aches cured, by Riding'. And in *Primitive Physic*, published continually between 1747 and 1859, John Wesley, who was said to ride well over 6,000 km a year, wrote that riding was a 'grand medicine' of particular value to those leading a sedentary life.[75] John Eliot usually spent three to four hours in the saddle while Thomas Bridge managed anything from two to five hours. Such a serious workout does not appear to have been unusual. One Sunday in April 1777, John Allen, a twenty-year-old brewer's son, completed a 61 km round trip from Wapping to South Mimms. It was not until Tuesday that he felt 'Recover'd from the fatigue' but Allen continued to relish every opportunity to borrow his father's horse 'to ride out'.[76] Whether at their books in counting houses, running manufactories, conversing in coffee houses or dining with friends, many middling Londoners were cooped up indoors for much of the week. Riding out provided energetic relief but at the same time, equine companionship provided a powerful alternative to the kind of intense human contact that made commercial life so stressful. Riding out after a day of networking and negotiating allowed businessmen to communicate without words and to have their wishes granted without protracted negotiation. It was perhaps for this reason that Eliot was almost as likely to ride out alone as he was in company, while Bridge rode solo on a quarter of his outings. Mounting a horse lifted the rider clear above the stresses and strains of commercial life and allowed him, as one riding master put it, to 'enjoy' himself in 'contemplation'.[77]

It is now widely accepted in societies across the world that riders and horses develop close personal bonds. Numerous qualities have been suggested to explain why people are drawn to horses. The ecologist Paul Shepard, for instance, highlights the sensuousness of their 'sleek' coats and curvaceous bodies. 'Close up, the horse makes the heart beat faster', a sensation which, Shepard argues, is multiplied by the pleasure of genital stimulation when riding. Alternatively, the anthrozoologist Lynda Birke observes that people often see horses as individuals, friends and 'partners on particular journeys' adding that they 'learn to read each other' and 'as partners, horses become almost people'.[78] Whatever their interpretative differences, most studies recognise that riding horses generates intense emotions. By contrast, historians looking at horsemanship in the early modern period have tended to focus on issues of control and discipline, often as a means of commenting on social power structures. In doing so, they generally concentrate on riding manuals. While valuable in many respects, this kind of source offers an unbalanced view of the relationship between horse and rider because rationalism and political metaphors obscure the physical and emotional complexities of everyday interactions.[79] Historians need to question how the edicts of equestrian manuals were interpreted, ignored and rejected by those who read them while also considering the emotions that equine companionship generated. This is no easy task but some Georgian riders did make intriguing comments on the subject. In 1837, the London coach-maker and part-time locomotive engineer William Bridges Adams leapt to the defence of horse riding as travel by steam train began to grow in popularity. Steam, Adams proclaimed

> is a mere labourer – a drudge. It is not so with a horse... They are beautiful and intelligent animals, powerful yet docile... The man who rides a horse, feels a pleasure when the creature responds willingly to his purposes; and when he responds unwillingly, he feels a pride in the exercise of his power to compel him to obedience. Even when a horse is vicious, there is a pleasurable excitement in riding him. The rider's nerves are strung, his senses are quickened; eye, hand, and ear are alike on the alert; the blood rushes through the veins, and every facility is aroused.[80]

Few surviving diaries or letters describe riding experiences in any great detail and it is even harder to glean a sense of how riders felt about their horses. A precious exception is the diary of Sylas Neville who, in the late 1760s, devoted himself to becoming an elegant metropolitan horseman. While taking tuition at Angelo's academy, Neville acquired a horse which he graced with the name Pizarro, after the heroic conquistador. He described this painstakingly selected steed as 'a very pretty dark bay' horse 'of a size fit to carry a light weight and not too high for me' and soon developed a remarkable affection for the animal. This came into sharp focus when Neville fell into debt. After riding alone to Yarmouth in October 1772, he admitted '[I] cried almost all the way over my poor horse, which I may perhaps never ride again'. Even more revealingly, in 1784, he lamented, 'I seldom ride on horseback as I do not keep horses of my own; those I kept were so good that I cannot ride a bad or even a tolerable horse with any satisfaction'.[81] While some attitudes to animals have changed significantly since the eighteenth century, then as now, many recreational riders viewed their horses as individuals and valued companions. While more forthcoming with his feelings than most, Neville was not unusual in deeply appreciating the company of horses. In 1768, Lady Mary Coke wrote to her sister, the Countess of Strafford, to lament the 'loss of Your Horses'. She agreed that 'a loss that can be repair'd with money is not a misfortune, yet to those who love riding, a favourite Horse is a bad thing to lose'. The time and money that equestrian Londoners invested in care for their horses evinces a strong commitment to their wellbeing and a complex emotional attachment to these animals. It is revealing that Thomas Bridge recorded in his diary that he regularly 'went to the stables to see the Horses' or to 'dress' them even when he had no plans to ride.[82] Horse racing reveals a little more about this relationship.

Unlike modern race-goers who interact as spectators and gamblers, Georgian Londoners were determined to participate with their horses.[83] The dividing line between racers and spectators was ill-defined and unguarded, and many race-goers considered themselves fellow jockeys.[84] Visiting Epsom in the 1760s, the Frenchman Pierre-Jean Grosley was shocked to find

> neither lifts nor barriers... the horses run in the midst of the crowd, who leaves only a space sufficient for them to pass through... The victor, when he has arrived at the goal, finds it a difficult matter to disengage himself from the crowd, who congratulate, caress, and embrace him.[85]

Unsurprisingly, such interaction resulted in frequent accidents. At Tothill Fields in 1736,

> a young Fellow being in Liquor, riding furiously about the Course, beat down a Girl of about nine years of age, and rode over her... [later] the same person, riding amongst the thickest of the people, beat down and trampled on a young lad... and broke one of his legs... The fellow rode clear off.[86]

The excitement generated by horse races appears to have encouraged rumbustious riding on the journey to and from the courses. Onlookers described returning race-goers with a mixture of wonder and horror as they hurtled into the city. In 1771, a critic of Barnet races described how these unskilful 'London jockies' 'press with the utmost eagerness, or rather madness, through the narrow passages at and near Barnet, and then down... Highgate-hill'.[87] Alarming for some and exhilarating for others, this behaviour reminds us how closely the pleasurable sensation of speed was connected to the horse in Georgian minds. This is underlined by the reminiscences of the essayist Thomas de Quincey about coach travel in the early nineteenth century. In the 1840s, he observed that

> The vital experience of the glad animal sensibilities made doubts impossible on the question of our speed; we heard our speed, we saw it, we felt

> it as a thrilling; and this speed was not the product of blind insensate agencies, that had no sympathy to give, but was incarnate in the fiery eyeballs of the noblest among brutes, in his dilated nostril, spasmodic muscles, and thunder-beating hoofs.[88]

Horses were the focal point of a leading department of metropolitan culture in the Georgian period which led to the construction of grand equestrian venues, permeated London's parks and hinterland, and influenced the behaviour of thousands of people. In certain respects, the city's equestrian recreations were distinctly urban in character, promoting luxurious consumption, fashionable display and sociability. But horses also enabled and encouraged Georgian Londoners to escape the city and explore the countryside. At the same time, horses lured some, men in particular, away from society by offering an intoxicating combination of privacy, liberation, companionship, exercise and excitement.

Up to this point, this book has focused on the various roles played by horses and livestock in manufacturing, trade, consumption and recreation, but London's success in these areas heightened its reliance on another species. The next and final chapter considers one of the most serious challenges facing Georgian London: property crime.

7

WATCHDOGS

> He always does bark at strangers... no body can get into the stable, without 'tis somebody the dog knows.
>
> Thomas Croon, Old Bailey testimony (4 April 1744)

> I live in the Eight-Bell-yard, St Giles's... between three and four in the morning, I heard my dog bark very much, in my gallery: I got up and opened the door, and saw the prisoner... take down the stockings... I laid hold of her arm.
>
> Theophilus Wright, Old Bailey testimony (10 July 1765)

> I heard the dog bark, and a call of 'Watch'. I ran out, and found the prisoner in Thompson's custody; I found the [roof] lead tied in a handkerchief...
>
> John Philpot, Old Bailey testimony (21 April 1819)[1]

The Old Bailey Proceedings provide a spine-chilling acoustic record of criminal London. Victims scream, windows smash, bones crack and fires roar; but of all the sounds that accompanied crime in this period, one of the most persistent and telling was barking. Dogs were ubiquitous in the Georgian metropolis – in 1775, one newspaper estimated that the city was home to 100,000, based on the assumption that there was one dog for every ten people.[2] Yet, while dogs have snatched attention in studies of early modern and Victorian England, historians have tended to focus on fictional beings or reduce real animals to metaphors in debates about middle class domesticity, humanitarianism, luxurious consumption and social problems. As a result, scholars have become fixated on two canine types: the nuisance stray and the cosseted lapdog; and this has encouraged the assumption that dogs were either a hindrance to or a distraction from police and prosperity.[3] England's 'useful' dogs have received much less attention and too clear-cut a distinction has been drawn between so-called working dogs and so-called pets. This chapter blurs the boundary, showing that in Georgian London yard dogs and house dogs of every kind performed an important service: protecting property from thieves. Drawing on newspaper reports, the Old Bailey Proceedings and modern canine behavioural science, this chapter reveals that dogs exerted significant influence in the city. As well as examining the role that these animals played in the metropolitan economy, it will unpack the ways in which canine recognition, trust, loyalty and aggression impacted on human behaviour, as well as law and policing in this period.

Dogs served people in many different capacities and contexts in Georgian England. Farmers, shepherds and drovers relied heavily on herding dogs, while thousands of retrievers, pointers and fox hounds were used in hunting. In some parts of the country, small turnspit dogs worked in the kitchens of wealthy households, powering meat spits by running inside a wheel; and some urban street-sellers harnessed dogs to draw small carts. In Georgian London, however, the most common and influential role was that of the watchdog. Why should this be of interest to urban or social historians? For one thing, examining the lives of these animals offers valuable insights into the use and control of urban space, particularly of houses, shops, warehouses and commercial yards. Previous studies of the city have highlighted the expansion of ticketed access to leisure venues; the agency of masters, mistresses and landladies in guarding access to domestic space; as well as the power wielded by beadles, overseers and other officials in facilitating and blocking admission to workhouses in this period.[4] This chapter builds on this work by showing that people were not the only guardians of space in the Georgian metropolis. In doing so, it asserts that watchdogs demonstrated agency in three different ways. First, these animals foiled criminal plans, influenced human behaviour and enabled human actors, including their masters, watchmen and other officers. Second, they exhibited a degree of intentionality and self-directed action; and third, they shaped, albeit unconsciously, the urban experience of thousands of people.[5]

Property crime was a serious concern in Georgian London and led to major changes in policing and punishment in this period. To understand the role that watchdogs played, we first need to consider the environment in which they were operating. Prior to the Metropolitan Police Act of 1829, responsibility for law enforcement rested with parishes and private individuals. Night watchmen appointed by vestries patrolled the streets on regular beats and constables were empowered to summon local residents to assist in the pursuit of suspects. At the same time, rewards offered by the state and by victims encouraged thief-takers to track down criminals, negotiate the return of stolen goods, as well as inform on and blackmail criminals. Pressure to improve policing in London increased in waves of anxiety, linked to economic and military circumstances, which eventually contributed to the establishment of the capital's first detective force at the rotation office in Bow Street in 1753.[6] This was a significant development but limited resources and reliance on corrupt thief-takers hampered progress. Meanwhile, frustration with the night watch intensified: in 1754, it was complained that the watch remained 'very insufficient, for want of stouter and better conducted Men'.[7] Such criticism was often exaggerated – the city's watchmen had considerable and increasing strength in numbers, became ever more tightly regulated and were often instrumental in the arrest and conviction of criminals – but their crime-fighting efforts were diluted by having to deal with the homeless and inebriated, among other challenges.[8] Thus, in 1797, the merchant-magistrate Patrick Colquhoun complained that watchmen were 'of little use' because burglars simply waited for them to go on 'their rounds, or off their stands' before 'conveying the plunder to the house of the Receiver'.[9] By then, cargo worth an estimated £500,000 was disappearing from the Port of London every year. Finally, the city took action. Financed by West India planters and merchants, Colquhoun co-founded the Thames River Police, London's first centrally directed armed police force; in 1800, the government gave its official backing and the newly named Marine Police Force continued to restrict crime on the river and docks. Elsewhere in the city, however, thieves continued to exploit persistent weaknesses.[10]

Throughout the Georgian period, therefore, householders and businesses felt unable to rely on the authorities to protect their property and were forced to take matters into their own hands. Wealthy Londoners fuelled demand for increasingly sophisticated locks, which they incorporated into ever more fastidious security regimes. Every evening, tradesmen, shopkeepers and heads of households would make a sweep of their properties to secure doors and windows with shutters, locks, bolts and chains. As an added precaution, families squirrelled away their most valuable possessions in boxes, trunks, drawers, compartments and closets, each locked and carefully concealed.[11] Yet, as newspaper and court records attest, these security measures were far from impenetrable. In March 1739, the metropolitan press reported that a gang of thieves had broken into the cellar of an apothecary on Tower Hill. Their point of entry, a window, had been 'well secured by two strong Iron bars' but these soon gave way. The *Common Sense* newspaper was keen to point out that the burglars had 'met with no Disturbance from the watch' and that it was only because the family had been 'alarm'd by the barking of a little Dog' that the villains 'were forced to make a precipitate Retreat… without their intended Booty'.[12]

This chapter draws on a further fifty-one newspaper reports identified by searching the Burney Collection online with the keywords 'dog' and 'barking', and then extracting those reports that relate to metropolitan dogs interrupting housebreakers and burglars, who were variously described as thieves, rogues and villains. The limitations of this approach, in particular the imperfect 'significant word' accuracy offered by Optical Character Recognition when searching pre-1900 newspapers, mean that this dataset is unlikely to be complete.[13] The earliest relevant report found was published in 1730 but it was only in the last four decades of the century that one or more reports appeared on a nearly annual basis.[14] This does not offer convincing evidence of increasing levels of property crime or of the growing involvement of dogs in crime prevention because it, at least partly, reflects the expansion of the metropolitan press, significant changes in crime reporting and London's growing population in this period. Just over half of the reports in the dataset were published in the 1760s and 1770s, years in which London's newspapers increasingly filled space recently allocated to military reports with accident and crime reports.[15] As several historians have suggested, however, the demobilisation of military forces in the early 1760s does also appear to have intensified competition in the metropolitan labour market and contributed to rising prosecutions for property crimes.[16]

In the second half of the eighteenth century, watchdogs appealed to a metropolitan press eager for stories of thwarted burglaries, particularly when they humiliated, in one fell swoop, the city's cowardly thieves and incompetent watchmen.[17] In 1764, the *Gazetteer* reported that some rogues had attempted to rob the shop of a toyman and jeweller opposite the Royal Exchange. They managed to cut a hole through the window shutters but were deterred from entering 'by a dog that was in it'. The *Gazetteer* demanded 'Where were the Cornhill watchmen that evening?'[18] The following year, the paper gloated that 'some rogues' had attempted to break into a shop near Bishopsgate but that 'being alarmed by the barking of a little dog, the fellows made off without their booty'.[19] And in 1797, the *London Pack* reported that the cellar window of the Bricklayer's Arms at Limehouse was 'attempted to be broke open, but prevented by the perpetual barking of a little dog; one of the thieves was taken, and carried before Justice Staples'.[20] In keeping with the kaleidoscopic blend of messages which, as Peter King observes, typified crime reporting in this period, such stories managed to reassure, entertain, amuse and frighten readers, often all at once.[21] More importantly, in the context of this study, these brief but action-packed accounts provide extraordinary insights into the role played by dogs in the city's struggle against property crime. Many contemporary readers assumed that newspapers distorted the truth and we should treat these narratives with no less caution, but even sceptics like Horace Walpole accepted that crime reports 'seldom fail to reach the outlines at least of incidents'.[22]

This chapter also focuses on twenty-eight criminal trials identified by searching the Old Bailey Proceedings Online with the keywords 'dog' and 'barking' in the period 1715–1830. The first of these trials took place in 1731 but an earlier example appeared in 1678, and three more followed in 1698.[23] The final trial considered in this chapter was heard in 1829, the year of the Metropolitan Police Act, but dogs would continue to alert Londoners to thieves for decades to come.[24] The Proceedings do not provide anything like a reliable indication of how frequently dogs alerted Londoners to thieves, or how this may have changed over time. One reason for this is that the Proceedings only record incidents where a thief was apprehended and prosecuted, but even when a case was brought to trial, only some witnesses would have thought to mention the involvement of dogs. Furthermore, although trial reports became longer over the Georgian period, they only ever provided partial transcripts of what was said. It is almost certain, therefore, that the actions of many dogs would have been viewed as relatively minor details and omitted in the interests of brevity.[25] Nevertheless, the trial reports in which these animals do feature, offer a window onto canine interactions with their masters, thieves, watchmen and other human actors. Dogs had, of course, guarded property in London long before the eighteenth century but the nature of the evidence extant from the Georgian period means that their contribution is illuminated in unprecedented detail. This investigation ends with the establishment of the Metropolitan Police Force, which marked a major shift in the city's fight against crime, but this is not to suggest that dogs suddenly became less important in helping to deter and apprehend housebreakers.

Eighteenth-century London is closely associated with moral panics about crime but Robert Shoemaker has convincingly argued that 'Londoners adopted individualised approaches to crime, and often resisted the efforts of "moral entrepreneurs" to induce anxiety'.[26] The ubiquity of dogs in Georgian London could be interpreted as evidence of a high level of anxiety about crime but we need to keep in mind that many of these animals, particularly house dogs, were acquired for companionship as well as for security. What is clear is that by keeping a dog, Londoners were able to take effective action to reduce their risk of becoming a victim of housebreaking and burglary. Dogs offered two kinds of protection: a deterrent and an alarm. Ferocious dogs posed a serious occupational hazard for thieves and aroused genuine fear. Evidence of dogs physically attacking intruders is rare in the Old Bailey Proceedings, partly because defendants were unlikely to implicate themselves by admitting to being bitten at a crime scene, but also because victims of burglaries had reason to be cagey as owners of ferocious dogs could be charged for breaking nuisance laws or sued for damages. In 1827, however, the court did hear that a large Newfoundland dog had attacked a gang after they barged into a respectable residence near Regent's Park at noon. One of the housebreakers, John Duxberry, received a deep laceration to one of his legs. Aided by his accomplices, Duxberry managed to overpower and tether the dog but how to silence one of the loudest barks in the city? Thinking quickly, the gang threw down some meat, distracting their assailant long enough to make off with cash, jewellery, silverware and clothes worth more than £100. Their comeuppance came a few weeks later when the sister of an absent gang member turned informant.[27] This was an unexpected turn of events but Duxberry was just as unlucky to have been bitten.

Recent animal behaviour studies have observed that guard dogs bark or snarl when confronted but generally back away or remain hidden to avoid being harmed.[28] Nevertheless, sensible burglars steered clear of properties known to keep vicious animals or made plans to avoid confrontation. In 1818, the Old Bailey heard that Stephen Morris, a watchman-turned-thief, had broken into the Camden residence of William Clulow and snatched silverware valued at £45. Having previously apprehended thieves in the area, Morris not only knew that Clulow owned a dog but also that it had a propensity to bite. Morris' solution was to pay Joseph Braid, the family's footman, to shut 'the dog up in the scullery' before letting him into the house, but not all thieves could count on the help of an insider. Braid later testified that Morris had been afraid the dog 'should bite him', but in Georgian London, a dog's bark was almost always more deadly than its bite, triggering a chain of events that often led to arrest and trial, frequently followed by death or transportation.[29] It is, however, the story of a lucky escape which provides some of the most interesting detail about the contribution made by watchdogs.

In 1784, the *Whitehall Evening Post* reported with evident glee that a gang breaking into the Goat Alehouse in St John's Street had been 'attacked by a large mastiff dog, who pursued the villains to some distance from the house, and at length seized one of them'. Their screams attracted the night watchmen but not realising 'the faithful part the dog had acted', they 'rescued the villain out of his custody'. As well as mocking the failure of the watchmen, this story sought to impress readers with the fact that the dog then went home and 'kept incessantly barking til he had alarmed the family'.[30] This detail is significant because it evinces canine intentionality and self-directed action. By finding its way back home, 'some distance away', the mastiff proved that it could navigate via landmarks and form mental maps, even in a busy urban environment. But even more remarkable is the observation that instead of continuing to bark in the street to no end, the dog made the decision to return to its master and alert the family. As Chris Pearson has noted, modern studies of canine psychology identify such behaviours as evidence that dogs 'are not governed solely by instinct as they are able to deploy strategies to get what they want or need within relationships'.[31] What makes this source so significant is the newspaper's confidence that its readers would recognise this behaviour as proof of canine sagacity. Ten years earlier, Oliver Goldsmith had observed that a dog 'either conquers alone, or alarms those who have most interest in coming to his assistance', and presented this as confirmation that the dog was 'the most intelligent of all known quadrupeds, and the acknowledged friend of mankind'.[32] By the final quarter of the eighteenth century, newspapers, natural histories, sporting literature and children's books commonly ranked the 'sagacity' and 'nobility' of dogs above that of all other beings except man, horses and elephants, based on their ability and willingness to serve humans. Thus, in the 1790s, Ebenezer Sibly asserted that the 'sagacity and intelligence' of dogs 'bespeak an intellect, which by culture and education might be made subservient to the domestic conveniences of man'.[33] Nowhere in Georgian England exploited this knowledge more than London.

Watchdogs

The city's dogs provided a valuable free service to parish rate-payers because their watchmen were often drawn to crime scenes by the sound of barking. Early one morning in 1775, for instance, 'the continual barking of a butcher's dog' in the Borough 'alarmed the next watchman' to a break-in at a pawnbroker's shop. On the watchman's approach, 'the rogues made off', leaving behind a chisel, a hammer and a bunch of keys.[34] In more exceptional circumstances, vestries employed watchdogs to guard a vulnerable site and thereby free the watchmen to go about their rounds. In 1767, the vestry of St George, Hanover Square, stationed 'a large mastiff dog' in the burying ground after 'having been lately robbed of several dead bodies'. The plan proved a spectacular failure when a gang of body-snatchers somehow managed to steal both the dog and a corpse.[35] Nevertheless, it is clear that dogs were valued not just by residents but also by a governing elite which was determined to strengthen law and order in the metropolis.

The watchdogs of Georgian London were a motley assortment of lapdogs, terriers, hunting dogs, mastiffs and mongrels but what these animals had in common were their acute senses of sight, hearing and smell, combined with a powerful instinct to bark at perceived threats. Recent animal behaviour research has shown that even when sleeping, dogs make much more vigilant guards than their masters. Humans sleep for several hours in a single session during which we approach consciousness but generally avoid waking up. Canine sleep cycles operate very differently: a study conducted in the 1990s showed that a sample of domestic dogs in one Australian city slept, on average, for just sixteen minutes before spontaneously waking up. These animals remained alert for around five minutes before settling back down to sleep.[36] A. Roger Ekirch has convincingly argued that 'Until the modern era, up to an hour or more of quiet wakefulness midway through the night interrupted the rest of most Western Europeans' and presents a modest amount of evidence for Georgian London.[37] This kind of segmented sleep almost certainly gave some Londoners a better chance of combatting intruders, particularly if they arrived at or soon after midnight, but the evidence presented here shows that it did not diminish the city's reliance on dogs, which awoke far more frequently throughout the night. This is significant because thieves often chose to break into houses in the early hours of the morning when, as William Blackstone put it, sleep had 'disarmed the owner, and rendered his castle defenceless'.[38]

Georgian dog owners recognised and appreciated the near constant vigilance demonstrated by their animals. It is telling that when Oliver Goldsmith, a Londoner from 1756 to 1774, extolled the virtues of dogs in his *History of the Earth, and Animated Nature* (1774), he began by praising their ability to watch over the home:

> When at night the guard of the house is committed to his care, he seems proud of the charge; he continues a watchful centinel, he goes his rounds, scents strangers at a distance, and gives them warning of his being upon duty. If they attempt to break in upon his territories, he becomes more fierce, flies at them, threatens, fights, and either conquers alone, or alarms those who have most interest in coming to his assistance.[39]

And almost fifty years later, *The Sportsman's Repository* (1820) was similarly impressed that a dog 'would continue pacing backward and forward, marching and countermarching, with all the regularity of a sentinel throughout the night'. To illustrate this point, the book featured an engraving of a mastiff guarding a metropolitan timber yard (see Figure 33), to which we will return.[40]

Watchdogs did not have to be large or powerful to prove their worth. According to *The Sportsman's Repository*, Sir John Fielding, the chief magistrate at Bow Street from 1754 to 1780, had heard from thieves that 'they never dreaded half so much the attacks of the fiercest large dog, as the tongues of the smallest, which they could find no possible means to quiet, but knocking them on the head'. Little dogs were thought to sleep less heavily and the *Repository*'s author, John Lawrence, asserted that they 'may even be said to be watchful in their sleep'.[41] This complicates the impression generated by numerous eighteenth-century commentators, and repeated by historians, that small dogs were the very embodiment of idle luxury. In the late eighteenth century, caricaturists frequently depicted effeminate macaronis prancing through the West End with their Whippets and women fussing over lapdogs (see Figure 34) because,

33 John Scott after Philip Reinagle, *Mastiff*, engraving published in [John Lawrence], *The Sportsman's Repository; Comprising a Series of Highly-Finished Engravings, Representing the Horse and the Dog*... (1820).

as Ingrid Tague has argued, 'pets embodied the worst excesses of fashionable consumption, thanks to the fact that in addition to their status as fashionable goods, they were also literally consumers'.[42] Yet, as we have already seen, other strands of eighteenth-century discourse presented dogs as some of the most useful and noble of animals. Goldsmith wrote that a dog was 'Always assiduous in serving his master, and only a friend to his friends' but he was also keen to point out that there were tasks which new imported breeds were unsuited for, complaining, for instance, that 'large Dane' dogs 'are chiefly used rather for shew than service, being neither good in the yard nor the field'. Nevertheless, Goldsmith maintained that a wide range of dogs, including 'mongrels, of no certain shape', could usefully 'alarm the family' and that the only 'perfectly useless' animals were 'lap-dogs'.[43] On this last point, Goldsmith was mistaken, because no matter how artificially their owners behaved, lapdogs never lost their natural instincts and with the right provocation would

34 Anon., *Lady Nightcap at Breakfast* (hand-coloured mezzotint, 1772).

bark at and perhaps even bite an intruder. Thieves were unlikely to fear their tiny jaws but understood the threat posed by their shrill yapping.

Small dogs, including lapdogs, were more than capable of raising the alarm and newspaper reports often pointed out that it had been 'the barking of a little Dog' which had prevented thieves from robbing Londoners in their sleep. These reports did not specify breeds and while some small dogs would have been terriers and mongrels, the increasing ubiquity of Maltese Shock Dogs, Pugs and King Charles Spaniels in the eighteenth century meant that these fashionable lapdogs must have contributed something to domestic security. Tague is right to argue that the form and meaning of pet-keeping varies significantly between periods and cultures, but perhaps makes too clear-cut a distinction between 'working' dogs, which lived outside in a kennel or a yard, and 'pets', which shared domestic space with their owner, their 'primary purpose' being 'entertainment and companionship'.[44] While some of London's yard dogs conceivably offered companionship and entertainment, as well as keeping watch, these animals were viewed, first and foremost, as working dogs. But evidence from Georgian London also shows that house dogs were expected to raise the alarm in the event of a break-in or fire, and this purpose was just as primary as being a companion or a source of entertainment. Irrespective of how pampered a pet became, it never stopped being a watchdog. For this reason, I would argue that the increasing acceptability of pet-keeping in the eighteenth century was not only due to the rise of sentimental attitudes but also to an increasing reliance on canine security.

A broad spectrum of Georgian Londoners kept watchdogs. At the lower end of the property-owning ladder, artisans and shopkeepers found that they were more affordable than locks and no less effective. In 1777, 'the violent Barking' of a dog belonging to a stay-maker in Goswell Street 'alarmed the Family' and despite having removed the iron bars from the cellar window, the villains were forced 'to decamp without their Booty'. And on a Saturday night in 1783, 'the barking of a small Dog' interrupted a gang as they were about to break down the back door of a button-maker's residence in Upper Moorfields.[45] Yet, even wealthy Londoners, who could afford servants and locks, benefited from keeping a vigilant dog in the garden, hallway or under the bed. In the 1790s, a Bow Street Runner told the Old Bailey that the affluent Haynes family of Great Russell Street kept a 'tarrier dog' which 'made a barking' if 'any strange person' entered.[46] And a few years earlier, a dog thwarted the attempts of four men to break into the even grander residence of Viscount Stormont in Portland Place. The dog started barking when it noticed an unfamiliar figure lurking outside. This aroused the suspicion of Stormont's servants but all remained calm until one o'clock in the morning 'when on his barking again, they immediately got up, and saw a man… climbing back again over the rails'. With the house now on high alert, the thieves slipped back into the night, none-the-richer.[47] A plethora of businesses also depended on canine surveillance and threat.

Suburban farmers were particularly vulnerable to thieves because policing was at its weakest in the outskirts of the city. Keeping dogs on-site helped to protect valuable livestock, grain and tools from itinerant thieves: in 1774, Goldsmith wrote that a farmer's dog 'conducts' livestock but also 'guards them… and their enemies he considers as his own'. The following year, an incident in Islington proved Goldsmith's point. On the night of 17 September, a dog belonging to the poulterer Charles Laycock barked as thieves attempted to steal his geese; hearing the alarm, one of Laycock's men 'ran down, and pursued them across the Field with the Dog but they got off'.[48] Across the city, dogs also served tailors, weavers, hosiers, grocers, pawnbrokers, butchers, carpenters, innkeepers, ironmongers, plumbers, wheelwrights, curriers, tanners (see Figure 5), and countless other trades which were threatened by thieves, day and night. It should come as no surprise, therefore, that an advert published in 1787 deemed a yard dog as integral to the business of a tallow chandler or a soap-maker as a draught horse, carts, a melting pan and a tallow press.[49]

We have no way of knowing how much property dogs saved from thieves but the Proceedings reveal that they were trusted to protect goods of great value. On 4 June 1778, a dog 'left loose in the yard' of Cripplegate's White Horse Inn was stationed to guard a warehouse containing more than £1,000 worth of silver plate. In the 1780s, Thomas Powell, a currier in Bunhill Row, locked a 'large' and 'highly esteemed' dog in his shed to guard dozens of valuable hides and seal skins; and in the 1820s, the Spitalfields silk weavers, Waterlow and Sawyer, left a dog in front of their warehouse door to protect merchandise worth hundreds, if not thousands of pounds.[50] It is important to note that because many tradesmen and shopkeepers lived above their workplaces, a well-positioned dog often protected a family's domestic and commercial capital simultaneously. In 1776, for instance, the *Morning Chronicle* reported that thieves had attempted to break into a broker's shop in Drury Lane 'but were prevented from effecting their purpose by the barking of a small dog which lay in the shop, and whose noise alarming the family, the villains made their escape'.[51] Such stories publicised the fact that dogs made an economic contribution to households and showed that they were fulfilling their 'God-given destiny' to serve humans.[52]

The need for canine deterrence was most acute in London's giant warehouses where, the Proceedings reveal, fierce dogs guarded dyes, spices, silks and calicoes, among other valuable imports. The East India Company almost certainly employed dogs although this is difficult to prove because swathes of its Committee of Warehouses papers were destroyed in the nineteenth century.[53] What we do know is that some of the capital's most formidable guard dogs belonged to carpenters and timber merchants. These businesses were vulnerable to thieves not only because so many people visited them during the day but also because it was often impossible to enclose yards with high walls on all sides. The presence of an imposing mastiff provided an effective deterrent against trespass. Thus, in 1769, the *Gazetteer* carried an advert for a dog said to be ideal for a timber yard because it was 'supposed to be the largest dog in England, though not two years old' and 'barks remarkably loud'.[54] A similarly impressive specimen appears in *The Sportsman's Repository* engraving mentioned earlier (see Figure 33). Muscular and large-headed, the animal can be seen pacing around a timber yard on the south bank of the Thames secured to a post or a kennel by a thick collar and chain. This detail indicates that the yard is still open for business because while owners of enclosed yards generally released their dogs to patrol freely at night, during working hours, they had to balance the animal's ability to intimidate, with the threat of legal action, an issue to which we will return. These animals helped to flush out thieves from crowds of carters and carpenters who had legitimate business to be in the yard. This ability was confirmed by a canine behavioural science study in the 1990s, which showed that dogs working in busy yards tend to react less aggressively to people entering their territory but also that they can differentiate between normal behaviour and the suspiciously furtive movement of a thief creeping around at night.[55]

Selective breeding over centuries may have honed these instinctive abilities and some dogs in Georgian London may have received specific training but social historians should resist the temptation to conceptualise watchdogs as tools programmed and wielded by humans. By the early eighteenth century, selective breeding had undoubtedly created dogs with particular physical attributes and behavioural characteristics but the significance of the latter should not be overstated. As the psychologist Stanley Coren asserts, 'Our genetic manipulations have only minor effects on *which* specific behaviours a dog uses, but have a strong effect on how *frequently* such behaviours are expressed... thus terriers bark at anything that raises their excitement level'.[56] It is also important to recognise that Georgian Londoners did not need to train their dogs to bark at intruders, they did so instinctively. Many owners, however, would have endeavoured to stop them barking on command and those who wanted their dogs to attack intruders may also have tried to teach them to bite particular parts of the body. William Ellis, a Hertfordshire farmer and agricultural writer, wrote of seeing dogs in England trained to immobilise men by biting their hand via their sticks, or by seizing their foot. These animals would, he explained, 'oblige him to stand still, till by the Word of his Master's Command, he was released'.[57] Sensible owners understood the importance of being able to call their animal off an intruder to avoid them causing serious injury or death, which, as we will see, could lead to legal action. These are significant examples of human manipulation of and control over canine behaviour but when watchdogs served the interests of their master, they still did so, to a large degree, self-directed and with their own interests firmly in mind. Dogs barked at and bit intruders because it was in their own interests to guard what they perceived to be their own territory, and to protect the people who fed and cared for them.

Thus far, this chapter has focused on demonstrating that dogs played a significant role in defending trade and commerce, the lifeblood of London's prosperity. I now want to consider their deeper influence, the possibility that canine behaviour shaped human behaviour and social relations in ways that inanimate locks and bolts could not, and in ways that sometimes ventured beyond the reach of human control.

No amount of training could make dogs bark at burglars alone: creaking floorboards, strangers talking in the street, wheels on cobbles and countless other sounds could set them off; and one dog's bark encouraged others to respond, triggering a chain reaction capable of stirring entire parishes. Incessant barking, especially at night, infuriated Georgian Londoners, as it had done for centuries. In 1660, Samuel Pepys complained that he had been so 'exceedingly disturbed in the night with the barking of a dog of one of our neighbours, that I could not sleep for an hour or two, I slept late; and then in the morning took physic'.[58] English nuisance law offered some protection against those who kennelled large numbers of dogs in a residential area but there was no recourse against the owner of a yapping house dog. Driven to their wits' end, some Londoners became vigilantes. In 1798, the Court of King's Bench heard that two men had hanged a dog 'complained of by many of the neighbours as a nuisance, by barking all night'. The dog's incensed owner took a successful action of trespass but in reaching his verdict, Lord Kenyon judged that the animal's relentless barking meant that it had not been hung 'maliciously' and awarded a paltry shilling in damages.[59]

The Sportsman's Repository observed that when there was any real cause for alarm, a dog's bark would become 'loud, sharp, and quickly repeated, a distinction of which those within doors were well aware, and by which they were effectually alarmed'.[60] Recent scientific studies confirm that the acoustic structure of barks varies predictably with context, meaning that a dog's bark becomes significantly harsher, lower in frequency, longer in duration and more rapidly repeated when the animal has been disturbed, as opposed to when it feels isolated or is playing.[61] Certain breeds and individual dogs are more prolific barkers than others, however, and in Georgian London, owners of less discriminating animals were more likely to ignore them. This could prove costly, as the publican of the White Horse in Spitalfields discovered on 28 September 1799. Between two and three o'clock in the morning, Thomas Hopwood 'heard the dog bark' in his tap room but opted to stay in bed; when he rose at four o'clock he 'found the back-window broke' and several items stolen, including clothes, a large copper and a goose. The thief was caught when officers from Worship Street in Shoreditch, one of seven new police offices established in London in 1792, discovered the stolen items in his accomplice's lodgings. In court, Hopwood's wife, Sarah, stated that both men had surveyed the house while drinking beer, which suggests that she set the officers on their trail.[62] This case emphasises that by the late eighteenth century, Londoners were utilising watchdogs alongside a range of other crime prevention and policing strategies, which included locks, watchmen, constables, police officers and thief-takers.

Some of the most remarkable insights provided by the Proceedings and newspaper reports relate to fleeting interactions between watchdogs and thieves. This evidence demonstrates that canine behaviour shaped the ways in which housebreakers went about their work, which is as significant in terms of agency as when dogs thwarted these criminals. Dogs posed a serious threat but they were not infallible and criminals developed a range of strategies to overcome their adversaries. The surest method was to stun or kill the animal by beating it over the head before it managed to raise the alarm. On 8 April 1814, a Bishopsgate tallow chandler found his watchdog 'in a state of stupefaction' and hardly able to 'stand upon his legs' after thieves broke into his warehouse and stole 327 kg of candles and 25 kg of soap.[63] Delivering the blow may have been brutally straightforward but creeping up on a dog undetected required stealth and a strong nerve. A botched attack was likely to enrage the animal and cause an unquellable commotion, as one gang discovered to its cost in Bloomsbury in 1774. At first, the grocer's house seemed an easy target – within minutes the thieves had prised open a sash window – but as the first of them crawled inside, a dog started to bark. Desperate to silence the animal, the men doused it with 'vitriol or aquafortis… and burnt him in a terrible manner', but much to the amazement of the *St James's Chronicle*, the thieves 'did not gain admittance, as the dog continued barking at them so that they were obliged to steer off'.[64]

The most premeditated and cowardly of ploys was to administer poison ahead of a break-in. The 'highly esteemed' currier's dog mentioned earlier died in agony on 16 November 1784 when a gang fed it butter laced with poison. The Old Bailey heard that having purchased an insufficient dose, the thieves were forced to wait two hours for the creature to collapse. William Astill was arrested when a witness to both the gang's plotting and the crime itself gave his name to a constable.[65] Thieves did not reserve their cruelty for menacing yard dogs, some were prepared to poison small house dogs which, as we have seen, were viewed as equally dangerous. On 5 July 1786, John Strong managed to steal £17 worth of silverware from the King and Keys in Fleet Street after poisoning the innkeeper's black and white spaniel which, he recalled, had 'been in the house above seven years'.[66] For many respectable Londoners, such incidents would have highlighted the spine-chilling cruelty of the city's low-born criminals but also the moral proximity between these predominantly plebeian men and four-legged brutes. Indeed, some newspaper reports gave readers the impression that dogs faithfully serving their masters demonstrated far greater nobility.[67] It would be too simplistic to suggest that dogs were weapons in a struggle between rich and poor because, as this chapter has shown, these animals served relatively humble Londoners as well as the city's elite. But these animals certainly stood between desperate men and potentially life-changing financial gain, while also threatening to condemn them to transportation or death.

Only the boldest thieves were prepared to confront dogs, as illustrated by an incident reported in 1751. Early one November morning, four men entered the yard of a cow-keeper in Marylebone Fields and his dogs promptly raised the alarm. Panicked, the thieves seized a servant and threatened him to keep the dogs quiet while they robbed the house. Confident that the dogs would remain steadfast, the servant refused. The gang carved off one of his ears but failed to enter the house.[68] This episode emphasises that crime prevention in Georgian London relied on more than human co-operation. It also demonstrates that a dog's ability to recognise and remember trusted people and to make the distinction between friends and foes made them a sophisticated security system. Other than loyalty, this was the quality for which dogs were most highly esteemed in Georgian Britain. Oliver Goldsmith enthused:

> He knows a beggar by his cloaths, by his voice, or his gestures, and forbids his approach... the dog is the only animal... who knows his master, and the friends of the family; the only one who instantly distinguishes a stranger.[69]

This instinctive canine ability empowered their masters but also prompted thieves to adapt their criminal strategies.[70] One of the greatest threats to a house, manufactory or warehouse was a disloyal employee and watchdogs were on the front line of a bitter struggle between their masters and larcenous workers. Because dogs learnt to trust familiar people, particularly when they treated them well, servants could gain privileged access to buildings and yards.[71] When a porter returned to his employer's warehouse in Lad Lane in 1774, for instance, his canine co-worker felt no need to bark. This might have saved the animal but an accomplice decided that it would raise the alarm when they got to work, so he crept up and slit its throat. As the creature lay dying, the thieves picked the lock on the warehouse door and carried off a cartload of stock.[72]

A familiar face was often enough to dissuade a dog from barking but cautious thieves called out the animal's name, stroked it and bribed it with food, as the following cases reveal. On 17 November 1769, a 'disorderly apprentice' released by a file cutter six months earlier, helped a gang to rob his master by calling his 'bull-bitch by her name' and keeping 'her quiet'. The live goose which the thieves hoped to steal proved less co-operative, however, and the master of the house heard 'her flutter very much' and he 'cried out – Stop thieves!'[73] On 14 September 1825, the silk weavers introduced above woke to find their warehouse door wide open and 500 yards of manufactured silk missing. In court, John Chappell, one of the company's apprentices testified that he had locked the warehouse door at dusk and checked that the firm's dog 'was in doors' as usual. Yet, no one heard the animal growl or bark all night because at ten o'clock, Samuel Crook, another apprentice, had called the dog to follow him and left the animal outside for the rest of the night. This left the coast clear for Crook's accomplice, Edward Mason, to break in. Both were found guilty and sentenced to death. Chappell was in no doubt about the importance of Crook's role, telling the Old Bailey that if he had not taken the dog out, the crime 'would not have been done'.[74] The final example comes from a trial for arson rather than theft but emphasises the extent to which canine judgement was trusted by masters and their servants. On 17 May 1829, James Butler managed to placate 'a very ferocious dog' belonging to George Downing, a Chelsea floor-cloth manufacturer, before setting fire to the manufactory. Downing told the Old Bailey that 'if any strangers came on the premises, the dog was very fierce, and barked at them' but recalled that he had seen the accused 'feed the dog and caress him'. Mercy Hill, Downing's servant, testified that 'the doghouse is by the stable-door; I can always hear the dog bark very distinctly when I am in the kitchen – it did not bark at all on the morning of the fire; I did not hear it'. The trial report suggests that this evidence was accorded considerable importance in court and, therefore, that the dog's behaviour helped to convict Butler, who was condemned to death.[75] Linked to this, the growing importance of watchdogs in London, as well as in other parts of the country, as trade and consumption expanded, also appears to have influenced changes in English law.

Dog owners were first legally compelled to compensate people bitten by their animals in the late ninth century but by the 1530s it had been established that liability required proof of prior knowledge of the animal's propensity to bite. Throughout the Georgian period, the outcomes of both civil and criminal cases continued to be guided by this principle. In criminal law, it was recognised that 'if a man hath a beast that is used to do mischief; and he; knowing it, suffers it to go abroad, and it kills a man;

even this is manslaughter in the owner: but if he had purposely turned it loose' it was murder. Nevertheless, throughout the eighteenth and nineteenth centuries, it remained extremely rare for owners to face criminal charges for assault, manslaughter or murder.[76] Civil actions brought for damages and cases based on nuisance law were more likely to succeed but even then, dog owners retained the upper hand. By the 1760s, English nuisance law had drawn a clear distinction between an unmuzzled dog being 'let loose' in a contained yard and one 'permitted to go at large'.[77] Thus, in 1796, the court of King's Bench refused to award damages to a carpenter's foreman attacked by the yard dog because the animal 'had been properly let loose, and the injury had arisen from the Plaintiff's own fault, in incautiously going into the Defendant's yard after it had been shut up'.[78]

A similar verdict was reached in the case of *Luccock v. Slater* in the Guildhall summary court two years later. Sir Lawrence Soulden judged that while Slater's 'large mastiff dog' had savaged several men 'in an alarming and dangerous way', it had been proven that 'when seized and pulled down' by the animal, the plaintiff had been 'walking over the premises of the Defendant'. Despite siding with the defendant, Soulden observed 'that it was very wrong for any man to keep so dangerous a dog' and that if the defendant did not hang it, he would make 'a Rule of Court that the dog should be hanged'. The *Sun* newspaper commented that this was

> a caution to any other person who may have a vicious dog, to destroy him on the first notice of any one having been bit by him; as the owner is afterwards liable to a prosecution and fine, or to an Action for damages.[79]

This earnest advice may have appealed to some middle class humanitarians but it would have fallen on mostly deaf ears. For many Londoners, owning a dangerous dog was well worth the risk to protect their property and families. The press took an interest in this case because it surfaced during a parliamentary debate about taxing dogs, a debate that reveals a great deal about contemporary attitudes to watchdogs and human–canine relations more generally.

Proposals to tax dogs had been heard periodically since the 1730s, with bills being debated in Parliament in 1755, 1761 and 1776, but it was not until 1796 that a tax was enacted and only then after fierce disagreement. Caricaturists had a field day. In *Effects of the Dog Tax*, Isaac Cruikshank depicted three of those who opposed the tax – Richard Brinsley Sheridan, Charles Fox and Charles Stanhope – as curs hanging from the gallows, having been judged 'Not worth the tax'. Six ministerial dogs watch them swing, including Prime Minister Pitt, who is shown chained to the Treasury kennel.[80] The bill was introduced on 5 April 1796 by John Dent, the MP for Lancaster, who complained that too many nuisance dogs were attacking livestock in the countryside. At a time of crop failures and food riots, many would have shared Dent's indignation that dogs were destroying valuable sheep and cattle while also consuming the offal and grain needed to feed the poor. The economic case for a dog tax claimed that it would reduce parish poor rates, lower food prices and provide Pitt's government with revenue to help defeat the French. At the same time, some viewed the tax as a way of ridding England of rabies – a disease that continued to provoke concern throughout the eighteenth century – by discouraging the poor from owning dogs and thereby reducing the number of strays roaming the country.[81] Others supported the tax as a way of tightening up the game laws and preventing the poor from poaching with dogs. Underlying both objectives, however, was a powerful bourgeois movement to reform the morals of plebeian men who, it was alleged, neglected both their work and their families.[82] In 1796, Edward Barry, a doctor and dog tax advocate, wrote that 'protection, and sport' were the only reasons for owning dogs and since the poor had little property to protect and no right to hunt, they had no justification for doing so. Others pointed out that dogs were a crippling luxury for the poor, as illustrated by George Morland's painting, *The Miseries of Idleness* (1790), which depicts a wastrel and his overdressed wife forcing their malnourished son to share a bone with the family's terrier.[83]

Among proponents of the dog tax, however, there was considerable disagreement about which dogs constituted useless and harmful luxuries. As early as 1740, William King had argued that 'Tanners, curriers, fellmongers, butchers, and the like' should be exempt because their trades required a dog, but Dent insisted that all dogs were luxuries and sought to levy 2s 6d on every animal, except guide dogs for the blind, in the expectation that this would lead to widespread culling.[84] Various arguments were used to oppose Dent's bill in Parliament. Some claimed that it was an insidious attempt to destroy the hunting privileges of all but the aristocracy, while John Courtenay and William Windham pointed out that dogs provided valued companionship to hard-working people and warned that the mass extermination of dogs would have a dangerous effect on the morals of the nation.[85] Arguably the most compelling speech, however, foregrounded the watchdog. Sheridan, the Whig MP for Stafford, began by ridiculing the very title of Dent's bill – which promised 'the protection of the persons and property of His Majesty's Subjects against the evil arising from the increase of dogs' – because, he pointed out, 'instead of supposing, as it generally had been supposed, that dogs were better than watchmen for the protection of property, people might be led to imagine that dogs were guilty of all the burglaries usually committed'.[86] The bill's intention to annihilate 'almost the whole species of Dogs' was, Sheridan went on, 'highly ungrateful; for they were not only our servants and friends, but they were our allies in the defence of law and order'.[87] In the same vein, a contemporary poem, *The Lamentation of a Dog, On the Tax, and its Consequences*, demanded 'Who watch the Labourer's Cloathes, the Waggon keep; / Protect the lonely Cot's few Hours of Sleep… / Dogs, by an old, poor, friendless Master's side, – / His safety dearer than their own – have died.'[88] Sheridan was not suggesting that the tax would prevent the rich from protecting their property but that labourers, artisans and shopkeepers would be made to suffer. To his mind, the security and companionship that dogs offered these families went hand-in-hand.

The government soon withdrew any support for Dent's bill and the dog tax which was enacted a few weeks later took a very different form, enshrining the principle that dogs owned by the rich were luxuries while those kept by the poor, whether for work or companionship, were necessities. Those who kept dogs which assessors deemed could be used for sport – greyhounds, hounds, pointers, setting dogs, spaniels, lurchers and terriers – along with those who owned two or more dogs, were charged 5s per animal. Households in possession of a single non-hunting dog were expected to pay 3s but only where they were already subject to the duties on inhabited houses or windows, which meant that a large proportion of plebeian dog owners were exempted.[89] This makes the numbers of animals in surviving tax rolls all the more impressive, although it has to be acknowledged that these came from affluent parishes where exemptions would have been relatively less common. In 1798, the tax was levied on about 400 dogs in the parish of St Martin-in-the-Fields, an area containing around 5,000 houses; while the similarly populous parish of St Margaret and St John produced about 700 taxable canines.[90]

While rural concerns stimulated much of the debate surrounding the dog tax, it also threw into sharp relief the essential role played by watchdogs in the metropolis at a time when the Industrial Revolution and global trade were filling the city with valuable goods and fuelling property crime in the process. These animals may often have been taken for granted, gaining little more than a pat on the head for thwarting thieves, but when Parliament questioned an Englishman's right to protect his property and to keep a faithful companion by his side, the dogs had their day. This chapter has shown that people were not the only guardians of urban space in Georgian London and that dogs of all shapes and sizes were instrumental in the protection of property, commerce and police. These animals blocked thieves and forced them to change their plans; they enabled a broad spectrum of Londoners to go about their business with greater confidence that their property was safe; and they assisted masters, watchmen, constables and others in detecting and apprehending criminals. All of this meant that dogs had a significant influence on the ways in which Londoners behaved, used urban space and interacted with one another and with them. This was often an unintentional influence but one that demonstrated agency nevertheless.

Conclusion

City of Beasts offers a reappraisal of what Georgian London was, how the city functioned and what role it played in some of the major developments of the period, including the agricultural, consumer and industrial revolutions. It goes about this by integrating animals into urban, social, economic and cultural history while simultaneously privileging evidence of lived experiences and tangible interactions between people and animals. The preceding chapters have shown that horses, cattle, sheep, pigs and dogs became increasingly ubiquitous in the eighteenth- and early nineteenth-century metropolis. But much more than this, they reveal that animals were with Georgian Londoners at the centre of their world.[1]

Pursuing beasts through the archives has led this book into parts of the city where historians have rarely ventured. In Tothill Fields, Turnmill Brook and Southwark, for instance, evidence of cows and pigs guided us to thriving agro-industrial sites involving brewing, distilling and animal husbandry. Thus, this book has examined unfamiliar dimensions to the city's physical, economic and cultural development. By tracing the work performed by the city's mill and draught horses, I have shown that Georgian London was industrious and productive in many more ways than historians have previously acknowledged. And by recognising the impact that horse mills had in brewing, tanning, brick-making, paint production and water supply, among other industries, this study makes it clear that the Industrial Revolution was not 'a storm that passed over London and broke elsewhere', but a process in which London played a central role.[2] Likewise, the contribution made by draught horses and their human co-workers highlighted in this book emphasises that London's role in domestic and global trade was not restricted to the port, Royal Exchange, counting houses, insurance brokerage offices and coffee houses. We have seen that the business of distributing goods dominated and at times overwhelmed the streets of the metropolis. Furthermore, evidence of milch cows and pigs being raised in the city, as well as the trades in meat on the hoof and horses, demonstrate that London was at the heart of developments in the agricultural and consumer revolutions, both as a voracious consumer and as a producer. This book does not offer a history of London's environmental or ecological impact in this period, although such a study is long overdue. It does, however, emphasise that the city exerted a powerful draw on the human and animal resources of the British Isles, as well as its crops and coal. In doing so, it serves as a reminder that the boundary between London and its hinterland involved the constant exchange of lives, commodities, knowledge and customs.

Opening our eyes to the city of beasts complicates orthodox assumptions about metropolitan culture in this period and challenges the focus which historians have placed on sociability and indoor recreational venues such as assembly rooms, theatres and coffee houses. We have seen that London was the focal point of a British equestrian obsession and that while horses facilitated sociability, they also posed an alluring alternative to human company. Glimpses of these interactions emphasise that Londoners crossed the boundary between urban and rural far more than is usually suggested and that pleasure was sought in a more-than-human sphere. Finally, recognising the agency of animals has enabled this book to challenge the notion that Londoners were passive victims of overwhelming levels of property crime in the Georgian period. We have seen that householders, shopkeepers, tradesmen and merchants took a determined and often effective stand against thieves by entrusting dogs to watch over private space and property. In its totality, therefore, this book has been empowered by animals to complicate some of the central assumptions about the social, economic and cultural life of London in the eighteenth and early nineteenth centuries.

This book did not set out to tell the story of animals in isolation but to consider their interactions with Georgian Londoners. In doing so, it has shed light on the lives of previously overlooked or poorly understood social types. These have included low-paid occupational groups such as carters, drovers, coachmen and grooms, but also prosperous mercantile men, tradespeople and shopkeepers. We have seen that horses, cows and pigs gave entrepreneurial Londoners valuable opportunities to supplement income, switch trades, leave service or expand commercial operations in a seasonal and unpredictable economy. In myriad ways, animals empowered Londoners, including those who were near the bottom of the heap. Drovers and carmen commanded large, powerful and potentially destructive beasts, a relationship that gave these men the power to assert themselves on the streets and bully other road users, including those from the city's elite. Meanwhile, mews servants used their privileged access to elite equipages and their acute horse sense to exploit the trust of masters, profit from the wider horse economy, seek promotion in service or use their skills to become self-employed. Animals were, therefore, facilitators of social mobility and contributed to the city's prosperity, but in doing so they also generated a great deal of social tension. Disgruntled neighbours presented pig- and cow-keepers for creating public nuisances; drovers, carters and coachmen were described as two-legged brutes for their conduct on the streets; and horse ownership fuelled envy, distrust and dispute.

This book reveals that there was a powerful class dimension to human–animal interactions in the Georgian metropolis. In many ways, animals benefited propertied Londoners more than its poor: middling and elite families were the largest owners of London's working horses and livestock, and gained most financially from the city of beasts. Fine riding and carriage horses played a powerful role in elite display and recreation, thereby re-enforcing distinctions of rank and wealth; and dogs guarded the property of those who could afford it, while helping to condemn typically low-born thieves to arrest, often followed by transportation or death. Meanwhile, plebeian workers and horses were bound together in increasingly strenuous work regimes. In his *Second Stage of Cruelty*, William Hogarth depicted a coachman maliciously whipping his horse's head, having worked the animal until it collapsed. The artist also condemns a snarling drover for beating a sheep to death and portrays a drayman as lazy, negligent and uncaring. This was a convenient assessment for middle class reformers at the time and historians have generally accepted Hogarth's diagnosis of unthinking plebeian brutality at face value but it does not adequately explain why those who worked with animals might have behaved in this way. By focusing on evidence of lived experiences and tangible interactions, this book has shown that urbanisation and industrialisation denied drovers, coachmen and carmen the conditions upon which the co-operation of their animals depended. A blood-curdling accident reported in the *London Chronicle* in October 1820 emphasises what an incongruous environment London had become for such a large convergence of animals. A bullock running down the Minories from Whitechapel, we are told, charged at 'several poor women' sat at their street stalls, leaving some 'much injured'. The enraged animal then ran through a court in Rosemary Lane, where it 'came in contact with a horse drawing a cart'; horns were plunged into the horse's belly and 'lacerated it in such a manner as to expose its entrails'. Finally, as the horse plunged backwards, a porter was killed 'by being jammed between the cart and a house'.[3] As I suggested in Chapters 2 and 4, road traffic accidents and acts of violence against animals often reflected the desperation of individuals who were working in poverty as well as in exhausting and perilous conditions.

Some readers may be disappointed that this book does not discuss in more detail, or more directly, the attitudes and emotions which Georgian Londoners held towards animals. It was never my intention to do so, partly because so much ink has already been spilled on the topic but also because so little has previously been written about tangible interactions between people and animals, or the contribution which this made to the social, economic and cultural life of the metropolis. To combine both topics in a relatively short study would have been foolhardy but I hope that this book does contribute something to what has become a fierce debate about attitudes towards cruelty and kindness to animals in the eighteenth and nineteenth centuries. In the 1980s and 1990s, historians such as Diana Donald and Hilda Kean suggested that sympathetic attitudes towards animals arose in the late eighteenth century because urban dwellers were becoming increasingly familiar with the cruelty suffered by horses and livestock on their streets.[4] Since then, however, historians have begun to question the notion that opposition to animal cruelty increased in the Georgian period at all. Rob Boddice has argued that 'there emerged no rigorous concept of animal rights' and that 'the principal concern' of the animal protection movement was 'the well-being of men'. Similarly, Emma Griffin concludes that there is insufficient evidence to suggest that 'philosophical and theological reflections penetrated deep into English society'.[5] The behaviour and attitudes of the many Londoners encountered in this book support these conclusions, but it is worth revisiting the 1828 parliamentary report into Smithfield Market one last time.[6]

The 1820s were years in which both the treatment of livestock and the organisation of Smithfield Market received unprecedented attention from Parliament and middle class reformers. The Cruel Treatment of Cattle Act was passed in 1822 to 'prevent the cruel and improper Treatment' of horses, cattle and sheep. Two years later, the Society for the Prevention of Cruelty to Animals (SPCA) was founded and for several years its members focused on enforcing the legislation. Historians have often viewed these developments as the culmination of growing compassion for animals but the Smithfield report undermines this assessment.[7] The committee, which ultimately rejected proposals to remove the livestock market from Smithfield, recorded the testimonies of seventy-eight witnesses. Just over a third worked directly with livestock, mostly as slaughterers, butchers, graziers or salesmen; the rest included tradesmen, shopkeepers, Corporation of London professionals, surgeons, engineers, bankers, land agents, clergymen and gentlemen.[8] While the 1828 report certainly does not provide a comprehensive view of metropolitan attitudes – no women were interviewed and plebeian witnesses are heavily underrepresented – it does allow us to test certain assumptions. Around half of the committee's witnesses responded to questions about cruelty to animals but only three – a surgeon member of the SPCA, a City bill broker, and an army officer on half pay – expressed this as a primary concern. These individuals complained that 'drovers inflicted cruelty in a horrible way' and that the slaughtering process was done 'in any bungling cruel manner' but they also expressed concern for the order of the streets and public safety.[9] Nearly a quarter of witnesses expressed some concern for animal suffering but appeared more anxious about nuisance, public safety and commercial interests. These complaints included bullocks breaking shop windows, frightening ladies and creating traffic jams; the loss of business caused by fear of cattle; drovers drinking and swearing; the deterioration of meat caused by drovers beating cattle; and blood from slaughterhouses dirtying the streets. Remarkably, a quarter of witnesses complained about these issues exclusively and expressed no concern for animal suffering. Thus, the report strongly suggests that even at the end of the Georgian period, the issue of cruelty to animals lacked sufficient public interest to stand up on its own two legs. This is not to say, however, that Londoners condoned all instances of cruelty when they encountered them.

One of the most striking aspects of the report is the tendency for committee members and witnesses to draw a distinction between 'necessary' and 'wanton' cruelty. This jars with conceptions of cruelty to animals in twenty-first-century Britain but is central to understanding attitudes at the time. Samuel Johnson's mid-eighteenth-century definition of 'Cruelty' as 'Inhumanity; savageness; barbarity' left plenty of room for interpretation. Likewise, Johnson defined 'wanton' as 'Loose; unrestrained'; 'wantonly' as 'carelessly'; and 'wantonness' as 'negligence of restraint'.[10] Thus, to accuse an individual of 'wanton cruelty' was to imply that, through carelessness or negligence of restraint, they had exceeded the parameters of 'necessary cruelty'. Of the twenty-nine witnesses who expressed a view about the cruelty practised in the livestock trade, seventeen described it as wholly 'necessary', while twelve described it as 'necessary' in some instances and 'wanton' in others. No witnesses were prepared to describe Smithfield's cruelty as wholly 'wanton'. While Smithfield cattle and sheep may have suffered the worst extremes of this way of thinking, it also clearly influenced the harsh treatment of working horses and dogs in the metropolis. Indeed, there was a broad acceptance that an ill-defined and unstable degree of cruelty or violence was justified in the management of all animals in this period.

Georgian London may not have witnessed the flowering of a new compassion for animals but I also want to dispel the pervasive myth that the Victorians somehow invented affection for animals. This book has shed light on numerous instances in which animals received appreciation, kindness and affection in the eighteenth century. We saw this in the enthusiasm and dedication with which wealthy Londoners rode and attended to their horses in Chapter 6; as well as in the behaviour of domestic pig-keepers, discussed in Chapter 3. Further evidence can be found in the paint and ink used to venerate the skill, strength and hard work of dray horses; and in the praise given to faithful dogs by the metropolitan press. If we want to gain a better understanding of how people viewed, felt about and treated animals in this period, we must begin by paying far greater attention to evidence of lived experiences. Only then is it possible to appreciate the contexts and temporalities of human interactions with animals. The resulting impression undermines simple narratives of improvement or decline.

It is tempting to assume that the treatment of animals in modern Britain has improved dramatically and in every respect since the eighteenth century, but this flatters our behaviour and misrepresents the actions of Georgian Londoners. Historians have condemned them for being cruel to animals but if this was the case, we also have to acknowledge that these urban dwellers looked their victims in the eye. By contrast, twenty-first-century Londoners largely consume animals which have been farmed and slaughtered out of sight and out of mind. This book has not sought to pinpoint a particular moment at which London's relationship with animals fundamentally changed but it is important to emphasise that the city of beasts long outlived George IV. While the employment of mill horses fell into sharp decline in the first quarter of the nineteenth century, the number of draught horses working in the city continued to increase throughout Victoria's reign.[11] Cattle, sheep and pigs did become much less visible in the city after 1855, the combined effect of the livestock trade's removal to suburban Islington, a sanitarian clampdown on working class pig-keeping and the rise of the dead meat trade facilitated by the expansion of the railways. Londoners continued to come into contact with livestock well into the twentieth century but not with anything like the frequency or intensity that their Georgian predecessors had done.

The story of a metropolis at the dawn of the modern age, which was both agriculturally productive and heavily horse-powered, matters in the twenty-first century. In 2000, Londoners consumed an estimated 385,000 tonnes of meat, 764,000 tonnes of milk and cream, and 63,000 tonnes of egg. But here, as in many other major cities in the West, animal husbandry is virtually non-existent.[12] To satisfy demand, London harvests the produce of millions of invisible livestock, animals which have never lived in the Greater London area. The social and cultural effects of this occasionally spark public debate. In 2007, the UK media seized on research exposing the apparent ignorance of children asked about the origins of bacon, burgers and eggs. 'City children' fared worst, proving themselves half as likely as 'countryside kids' to know that beef burgers came from cows.[13] Such findings make some laugh and others despair but there can be no doubt that Londoners have never been more physically isolated from the animals which they consume. This has significant social as well as environmental implications. The 2013 horse meat scandal in Europe underlined the degree to which urban societies in the West have lost track of the biological and geographical origins of the meat on their plates. But we should not lose sight of the fact that urban farming continues on a huge scale in many different forms and contexts in lower-income countries.[14] In the 1990s, it was estimated that 800 million of the world's urban dwellers were food producers, 100 million of whom sold their surplus produce. For how long this will continue is difficult to tell. The geographer Peter Atkins has observed that cities such as Chennai, Dhaka and Amman exhibit 'a strong survival of rural functions' including animal husbandry, but adds that this mode of life has been, and continues to be, under threat from urbanisation, modernisation and globalisation.[15] In 2006, a UN report observed that while urban and suburban husbandry offer 'a quick fix for countries in rapid economic development with fast-growing urban centres', these activities are eventually forced out of the city towards 'feedcrop areas, or transport and trade hubs where feed is imported'.[16] Thus, these city dwellers may soon join Londoners in their relative alienation from livestock. Around the world, urban planners, architects, environmentalists, scientists and local communities are working to reintegrate nature to make cities more sustainable and healthier to live in but whether this can involve significant animal husbandry remains unclear.[17]

Many of the issues surrounding human–animal interactions discussed in this book are pertinent to debates about twenty-first-century Britain, not least where social discord is involved. Those who work with animals, for instance, particularly farmers, continue to clash with groups who do not. The former often complain that animal rights and conservation activists do not understand them, their operations or nature itself. Other divisive issues rooted in the Georgian period include fox hunting and dog ownership, both of which generate censure and hostility between classes, as well as between town and country. At the same time, to take a very different example, the idea that animal companionship offers mental and physical health benefits, of the kind sought by Georgian equestrians, is gaining widespread acceptance, although many city dwellers still struggle to grasp these opportunities.

This book has highlighted various forms of animal agency. Every animal which we have encountered was an agent on the basis that they made a difference to other actors and most significantly, in the context of this book, to Londoners. I have focused on the many ways in which the physicality, power, intelligence and behaviour of animals variously enabled and encouraged Londoners to work, move, innovate, compete, profit, protect and unwind. But we have seen animals influence human behaviour in other important respects. The multifaceted appeal of horses, for instance, lured riders away from sociability while canine behaviour forced thieves to alter or abandon their plans. Moreover, I hope to have demonstrated that Georgian London was shaped by more than human needs. The capital's elite horses, in particular, were voracious consumers of food, manufactured goods, architecture and human care, which made them powerful agents in the consumer revolution and the metropolitan economy, as well as in the city's culture, social relations and physical development. Finally, I have suggested that the city's watchdogs demonstrated a degree of intentionality and self-directed action, which added an extra dimension to their agency.

Above all, this book argues that non-human animals were intrinsic to London's emergence as the world's first modern metropolis, and that the city was continually reshaped by their interactions with its people. In doing so, it hopes to open up new lines of historical enquiry in the study of London and Britain, but also to encourage the integration of animals into social and urban history internationally.

NOTES

PREFACE

1 William Wordsworth, *The Prelude or Growth of a Poet's Mind (Text of 1805)*, ed. E. de Selincourt, corrected by S. Gill (Oxford: Oxford University Press, 1970), p. 109.

2 I am grateful to Captain Gregory Flynn and Eloise Maxwell for confirming this.

3 Johnson proclaimed that 'By seeing London' he had 'seen as much of life as the world can shew' and that the city's 'wonderful immensity' lay not in 'the showy evolutions of buildings' but in its 'multiplicity of human habitations'. See James Boswell, *The Journal of a Tour to the Hebrides, with Samuel Johnson, LL.D.* (Dublin, 1785), p. 379, and *The Life of Samuel Johnson, LL.D.* (1791; Ware: Wordsworth Editions, 2008), p. 215.

4 E. A. Wrigley, 'A simple model of London's importance in changing English society and economy 1650–1750', *Past and Present*, 37 (1967), 44; J. Landers, *Death and the Metropolis: Studies in the Demographic History of London, 1670–1830* (Cambridge: Cambridge University Press, 1993), p. 179; L. D. Schwarz, *London in the Age of Industrialisation: Entrepreneurs, Labour Force and Living Conditions, 1700–1850* (Cambridge: Cambridge University Press, 1992), pp. 125–8; T. Barker, 'London: A unique megalopolis?', in T. Barker and A. Sutcliffe (eds), *Megalopolis: The Giant City in History* (Basingstoke: Macmillan, 1993); P. Bairoch, J. Batou and P. Chèvre, *The Population of European Cities, 800–1850* (Geneva: Librairie Droz, 1988), p. 283.

5 T. Hitchcock, R. Shoemaker, C. Emsley, S. Howard and J. McLaughlin, *The Old Bailey Proceedings Online, 1674–1913*, www.oldbaileyonline.org, version 8.0, March 2018, hereafter OBP; T. Hitchcock, R. Shoemaker, S. Howard and J. McLaughlin, *London Lives, 1690–1800*, www.londonlives.org, version 2.0, March 2018, hereafter LL. Both accessed 14 January 2019.

6 P. Earle, *A City Full of People: Men and Women of London, 1650–1750* (London: Methuen, 1994); Daniel Defoe, *A Tour through the Whole Island of Great Britain*, ed. P. Rogers (1724–26; Harmondsworth: Penguin, 1986).
7 P. Clark (ed.), *The Cambridge Urban History of Britain. Volume 2: 1540–1840* (Cambridge: Cambridge University Press, 2000); S. Inwood, *A History of London* (London: Macmillan, 1998); P. Ackroyd, *London: The Biography* (London: Chatto and Windus, 2000); J. White, *London in the Eighteenth Century: A Great and Monstrous Thing* (London: Bodley Head, 2012).

Introduction

1 R. Drayton, *Nature's Government: Science, Imperial Britain, and the 'Improvement' of the World* (New Haven and London: Yale University Press, 2000), pp. 87, 119–20.
2 William Cobbett, *Rural Rides*, ed. I. Dyck (1830; Harmondsworth: Penguin, 2001).
3 D. Donald, *Picturing Animals in Britain, 1750–1850* (New Haven and London: Yale University Press, 2007), pp. 199–200 and '"Beastly sights": The treatment of animals as a moral theme in representations of London, c. 1820–1850', *Art History*, 22 (1999), 524–6; R. Paulson, *Hogarth Volume III: Art and Politics, 1750–1764* (Cambridge: Lutterworth, 1993), p. 27; P. Beirne, *Hogarth's Art of Animal Cruelty: Satire, Suffering and Pictorial Propaganda* (Houndmills: Palgrave Macmillan, 2015).
4 J. Berger, 'Why look at animals?' originally published in Berger, *About Looking* (London: Writers and Readers Publishing Cooperative, 1980), republished in Berger, *Why Look at Animals?* (London: Penguin, 2009), p. 12.
5 W. Cronon, *Nature's Metropolis: Chicago and the Great West* (New York and London: W. W. Norton, 1991); M. V. Melosi, 'The place of the city in environmental history', *Environmental History Review*, 17 (1993), 1–23; C. M. Rosen and J. A. Tarr, 'The importance of an urban perspective in environmental history', *Journal of Urban History*, 20:3 (1994), 299–310; J. Wolch and J. Emel (eds), *Animal Geographies: Place, Politics and Identity in the Nature–Culture Borderlands* (London: Verso, 1998); J. Wolch, 'Anima urbis', *Progress in Human Geography*, 26 (2002), 721–42; W. Klingle, *Emerald City: An Environmental History of Seattle* (New Haven: Yale University Press, 2007); S. Whatmore, *Hybrid Geographies: Natures, Cultures, Spaces* (London, Thousand Oaks and New Delhi: Sage Publications, 2002); A. C. Isenberg (ed.), *The Nature of Cities: Culture, Landscape, and Urban Space* (Rochester, NY: University of Rochester Press, 2006); C. McShane and J. A. Tarr, *The Horse in the*

City: Living Machines in the Nineteenth Century (Baltimore: Johns Hopkins University Press, 2007); A. N. Greene, *Horses at Work: Harnessing Power in Industrial America* (Cambridge, MA: Harvard University Press, 2008); S. Castonguay and M. Dagenais (eds), *Metropolitan Natures: Environmental Histories of Montreal* (Pittsburgh: University of Pittsburgh Press, 2011). A particularly important European study is D. Roche, *La Culture Équestre de l'Occident, XVIe – XIXe Siècle: l'Ombre du Cheval, Tome 1, Le Cheval Moteur, Essai sur l'Utilité Équestre* (Paris: Fayard, 2008).

6 R. E. Park, 'The city: Suggestions for the investigation of human behaviour in the urban environment', in R. E. Park, E. W. Burgess and R. D. McKenzie (eds), *The City* (1925; 4th edn, Chicago, IL: University of Chicago Press, 1967), pp. 1–46; L. Mumford, *The Culture of Cities* (New York: Harcourt, Brace, 1938) and *The City in History: Its Origins, its Transformations, and its Prospects* (New York: Harcourt, Brace, 1961).

7 H. Leitner and E. Sheppard, 'Unbounding critical geographic research on cities: The 1990s and beyond', *Urban Geography*, 24 (2003), 510–28. See also L. S. Bourne (ed.), *Internal Structure of the City: Readings on Urban Form, Growth, and Policy* (2nd edn, New York: Oxford University Press, 1982), pp. 29–35; S. J. Mandelbaum, 'Thinking about cities as systems: Reflections on the history of an idea', *Journal of Urban History*, 11 (1985), 139–50; S. Roberts, 'A critical evaluation of the city life cycle idea', *Urban Geography*, 12 (1991), 431–49; Melosi, 'The place of the city in environmental history', 10–11; R. Lindner, *The Reportage of Urban Culture: Robert Park and the Chicago School* (Cambridge: Cambridge University Press, 1996); E. W. Soja, *Postmetropolis: Critical Studies of Cities and Regions* (Oxford and Malden, MA: Blackwell, 2000), p. 86; N. Thrift and A. Amin, *Cities: Reimagining the Urban* (Cambridge: Polity Press, 2002); Isenberg (ed.), *The Nature of Cities*; D. R. Judd and D. Simpson (eds), *The City Revisited: Urban Theory from Chicago, Los Angeles and New York* (Minneapolis: University of Minnesota Press, 2011), pp. 5–6.

8 B. Braun, 'Environmental issues: Writing a more-than-human urban geography', *Progress in Human Geography*, 29 (2005), 637; S. P. Hays, 'From the history of the city to the history of the urbanized society', *Journal of Urban History*, 19 (1993), 3–4, 8. On the need to unbound London's history, see M. Pelling, 'Skirting the city? Disease, social change and divided households in the seventeenth century', and M. S. R. Jenner, 'From conduit community to commercial network? Water in London, 1500–1725', both in P. Griffiths and M. S. R. Jenner (eds), *Londinopolis: Essays in the Cultural and Social History of Early Modern London* (Manchester: Manchester University Press, 2000), pp. 154–75, 250–72; see also G. Maclean, D. Landry and J. P. Ward

(eds), *The Country and the City Revisited: England and the Politics of Culture, 1550–1850* (Cambridge: Cambridge University Press, 1999).

9 R. Porter, 'Enlightenment London and urbanity', in T. D. Hemming, E. Freeman and D. Meakin (eds), *The Secular City: Studies in the Enlightenment* (Exeter: University of Exeter Press, 1994), pp. 40–1. See also R. Porter, 'The urban and the rustic in Enlightenment London', in M. Teich, R. Porter and B. Gustafsson (eds), *Nature and Society in Historical Context* (Cambridge: Cambridge University Press, 1997), pp. 176–94.

10 Braun, 'Environmental issues', 635–7; N. Smith, 'The production of nature', in G. Robertson, M. Mash, L. Tickner, J. Bird, B. Curtis and T. Putnam (eds), *Future Natural: Nature, Science, Culture* (London and New York: Routledge, 1996), pp. 35–54; C. Katz, 'Whose nature, whose culture? Private productions of space and the "preservation" of nature', in B. Braun and N. Castree (eds), *Remaking Reality: Nature at the Millennium* (London and New York: Routledge, 1998), pp. 46–63.

11 D. George, *London Life in the Eighteenth Century* (1925; Harmondsworth: Penguin, 1966); G. Rudé, *Hanoverian London 1714–1808* (London: Secker and Warburg, 1971); P. Langford, *A Polite and Commercial People: England 1727–1783* (Oxford: Clarendon Press, 1989); R. Porter, *London: A Social History* (London: Hamish Hamilton, 1994); Schwarz, *London in the Age of Industrialisation*; Earle, *A City Full of People*; Clark (ed.), *The Cambridge Urban History of Britain. Volume 2*; H. Greig, *The Beau Monde: Fashionable Society in Georgian London* (Oxford: Oxford University Press, 2013).

12 OBP; LL. On the need to identify variegated topographies, see M. Ogborn, *Spaces of Modernity: London's Geographies 1680–1780* (New York: Guilford Press, 1998) and J. Boulton, *Neighbourhood and Society: A London Suburb in the Seventeenth Century* (Cambridge: Cambridge University Press, 1987).

13 K. Thomas, *Man and the Natural World: Changing Attitudes in England, 1500–1800* (1983; Harmondsworth: Penguin, 1984). See also H. Ritvo, 'History and animal studies', *Society and Animals*, 10 (2002), 403–6; E. Fudge, 'A left-handed blow: Writing the history of animals', in N. Rothfels (ed.), *Representing Animals* (Bloomington, IN: Indiana University Press, 2002), pp. 3–18.

14 Thomas, *Man and the Natural World*, pp. 181–3.

15 Defoe, *A Tour*; Earle, *A City Full of People*.

16 Donald, '"Beastly sights"', 514–16. See also H. Kean, *Animal Rights: Political and Social Change in Britain Since 1800* (London: Reaktion Books, 1998), p. 13; H. Ritvo, *The Animal Estate: The English and Other Creatures*

in the *Victorian Age* (Cambridge, MA: Harvard University Press, 1987), pp. 125–6; K. Tester, *Animals and Society: The Humanity of Animal Rights* (London: Routledge, 1991); R. A. Caras, *A Perfect Harmony: The Intertwining Lives of Animals and Humans Throughout History* (New York: Simon and Schuster, 1996); E. Fudge, *Animal* (London: Reaktion Books, 2002) and *Perceiving Animals: Humans and Beasts in Early Modern English Culture* (Urbana: University of Illinois Press, 2000); D. Perkins, *Romanticism and Animal Rights* (Cambridge: Cambridge University Press, 2003); L. Kalof, *Looking at Animals in Human History* (London: Reaktion Books, 2007); C. Li, 'A union of Christianity, humanity, and philanthropy: The Christian tradition and the prevention of cruelty to animals in nineteenth-century England', *Society and Animals*, 8 (2000), 265–85.

17 In addition to the works cited above, see Thomas, *Man and the Natural World*; E. Fudge, *Brutal Reasoning: Animals, Rationality, and Humanity in Early Modern England* (Ithaca, NY: Cornell University Press, 2006); J. Passmore, 'The treatment of animals', *Journal of the History of Ideas*, 36 (1975), 195–218; C. Kenyon-Jones, *Kindred Brutes: Animals in Romantic-Period Writing* (Aldershot: Ashgate, 2001); D. Denenholz Morse and M. A. Danahay (eds), *Victorian Animal Dreams: Representations of Animals in Victorian Literature and Culture* (Aldershot: Ashgate, 2007); I. H. Tague, 'Companions, servants, or slaves? Considering animals in eighteenth-century Britain', *Studies in Eighteenth-Century Culture*, 39 (2010), 111–30; Donald, *Picturing Animals in Britain*; F. Palmeri (ed.), *Humans and Other Animals in Eighteenth-Century British Culture: Representation, Hybridity, Ethics* (Aldershot: Ashgate, 2006); R. Boddice, *A History of Attitudes and Behaviours Toward Animals in Eighteenth- and Nineteenth-Century Britain: Anthropocentrism and the Emergence of Animals* (Lewiston, NY: Edwin Mellen Press, 2008); R. Preece, *Brute Souls, Happy Beasts and Evolution: The Historical Status of Animals* (Vancouver: University of British Columbia Press, 2005) and (ed.) *Awe for the Tiger, Love for the Lamb: A Chronicle of Sensibility to Animals* (Vancouver: University of British Columbia Press, 2002); J. L. Wyett, 'A horse is a horse… and more: some recent additions to early modern animal studies', *Journal for Early Modern Cultural Studies*, 10 (2010), 148–62. Important exceptions include Ritvo, *The Animal Estate*, p. 4 and N. Pemberton and M. Worboys, *Mad Dogs and Englishmen: Rabies in Britain, 1830–2000* (Basingstoke: Palgrave Macmillan, 2007).

18 Primary sources favoured by British animal studies include René Descartes, *A Discourse on the Method of Correctly Conducting One's Reason and Seeking Truth in the Sciences*, ed. and trans. I. Maclean (Oxford: Oxford University

Press, 2006); John Locke, *An Essay Concerning Human Understanding* (1690); Jeremy Bentham, *An Introduction to the Principles of Morals and Legislation* (1789; Oxford: Clarendon Press, 1907), ch. 18; Humphry Primatt, *A Dissertation on the Duty of Mercy and Sin of Cruelty to Brute Animals* (1776); John Wesley, *A Survey of the Wisdom of God in the Creation or a Compendium of Natural Philosophy*, 3 vols (1770) and 'Sermon LX: The general deliverance', in *The Works of John Wesley*, ed. Thomas Jackson, 14 vols (3rd edn, 1829), vol. 6, p. 248; Jonathan Swift, *Travels into Several Remote Nations of the World. In Four Parts. By Lemuel Guliver, First a Surgeon, and Then a Captain of Several Ships* (1726; amended 1735); Anna Laetitia Barbauld, *The Mouse's Petition*, first published in 1773; republished in W. McCarthy and E. Kraft (eds), *The Poems of Anna Letitia Barbauld* (Athens and London: University of Georgia Press, 1994), pp. 36–7; Lord Byron 'Inscription on the Monument of a Newfoundland Dog', written 1808, published in *Lord Byron: The Complete Poetical Works*, ed. J. J. McGann, 7 vols (Oxford: Clarendon Press, 1980–93), vol.1, pp. 224–25; Percy Bysshe Shelley, *A Vindication of Natural Diet*, first published in 1813, reproduced in *The Prose Works of Percy Bysshe Shelley*, ed. E. B. Murray (Oxford: Clarendon Press, 1993), vol. 1, pp. 75–91.

19 I. H. Tague, *Animal Companions: Pets and Social Change in Eighteenth-Century Britain* (Philadelphia, PA: The Pennsylvania State University Press, 2015), p. 6.

20 J. Tosh, 'The history of masculinity: an outdated concept?', in J. H. Arnold and S. Brady (eds) *What is Masculinity? Historical Dynamics from Antiquity to the Contemporary World* (Basingstoke: Palgrave Macmillan, 2011), p. 23.

21 T. Hitchcock, *Down and Out in Eighteenth-Century London* (London and New York: Hambledon and London, 2004), pp. 233, 239.

22 E. Cockayne, *Hubbub: Filth, Noise and Stench in England 1600–1770* (New Haven and London: Yale University Press, 2007), pp. 1, 107, 148, 166–72, 192–3, 213.

23 Hector Gavin, *Sanitary Ramblings: Being Sketches and Illustrations of Bethnal Green, a Type of the Condition of the Metropolis and Other Large Towns* (1848), p. 87.

24 G. Phillips, 'Writing horses into American Civil War history', *War in History*, 20 (2013), 160–81; C. Philo and C. Wilbert, 'Animal spaces, beastly places', in C. Philo and C. Wilbert (eds), *Animal Spaces, Beastly Places: New Geographies of Human–Animal Relations* (London and New York: Routledge, 2000), pp. 1–35; S. Swart, *Riding High: Horses, Humans and History in South Africa* (Johannesburg: Wits University Press, 2010); J. Hribal, 'Animals, agency, and class: Writing the history of animals from below', *Human Ecology Review*,

14 (2007), 101–12 and *Fear of the Animal Planet: The Hidden History of Animal Resistance* (Petrolia, CA: CounterPunch, 2010).

25 C. Pearson, 'Beyond "resistance": Rethinking nonhuman agency for a "more-than-human" world', *European Review of History*, 22 (2015), 710.

26 B. Latour, *Reassembling the Social: An Introduction to Actor-Network-Theory* (Oxford: Oxford University Press, 2005), p. 71. See also P. Joyce and T. Bennett, 'Material powers: Introduction', in T. Bennett and P. Joyce (eds), *Material Powers: Cultural Studies, History and the Material Turn* (London: Routledge, 2010), p. 4.

27 See, for instance, W. H. Sewell, 'A theory of structure: Duality, agency, and transformation', *American Journal of Sociology*, 98 (1992), 1–29.

28 Latour, *Reassembling the Social*, p. 72.

29 D. Edgerton, *The Shock of the Old: Technology and Global History Since 1900* (London: Profile Books, 2006), p. 9.

30 C. Plumb, *The Georgian Menagerie: Exotic Animals in Eighteenth-Century London* (London: IB Tauris, 2015); Tague, *Animal Companions*.

1 Mill horse

1 J. L. Hammond, 'The Industrial Revolution and discontent', *New Statesman* (21 March 1925); I. J. Prothero, *Artisans and Politics in Early Nineteenth-Century London: John Gast and his Times* (Folkestone: Dawson Publishing, 1979); G. S. Jones, *Outcast London: A Study in the Relationship Between Classes in Victorian Society* (Oxford: Clarendon Press, 1971); T. S. Ashton, *The Industrial Revolution, 1760–1830* (London: Oxford University Press, 1948); Schwarz, *London in the Age of Industrialisation*; M. Reed, 'London and its hinterland 1600–1800: The view from the provinces', in P. Clark and B. Lepetit (eds), *Capital Cities and Their Hinterlands in Early Modern Europe* (Aldershot: Scholar Press, 1996), pp. 55–6.

2 M. J. Daunton, 'Industry in London: Revisions and reflections', *London Journal*, 21 (1996), 1; D. Barnett, *London, Hub of the Industrial Revolution: A Revisionary History 1775–1825* (London: Tauris Academic Studies, 1998), pp. 1, 35; Porter, *London*, p. 187; K. Bruland, 'Industrialisation and technological change', in R. Floud and P. Johnson (eds), *The Cambridge Economic History of Modern Britain. Volume 1: Industrialisation, 1700–1860* (Cambridge: Cambridge University Press, 2004), pp. 130–2; M. J. Daunton, *Progress and Poverty: An Economic and Social History of Britain, 1700–1850* (Oxford: Oxford University Press, 1995), p. 324; P. Hudson (ed.), *Regions and Industries: A Perspective on the Industrial Revolution in Britain* (Cambridge: Cambridge University Press, 1989).

3 The areas in which most engines were bought were: Lancashire (fifty-nine engines), London (thirty-six), Yorkshire (twenty), Staffordshire (nineteen), Nottinghamshire (seventeen), Shropshire (sixteen) and Cheshire (fourteen). Barnett, *London, Hub of the Industrial Revolution*, pp. 35–6; Boulton & Watt Archive and the Matthew Boulton Papers from the Birmingham Central Library (microfilm, Adam Matthew Publications): Part 3: Engineering Drawings – Watt Engines of the Sun and Planet Type *c.* 1775–1802.

4 E. A. Wrigley, *Energy and the English Industrial Revolution* (Cambridge: Cambridge University Press, 2010); P. Mantoux, *The Industrial Revolution in the Eighteenth Century: An Outline of the Beginnings of the Modern Factory System in England*, revised edn., trans. M. Vernon (London: Jonathan Cape, 1928); D. Landes, *The Unbound Prometheus: Technological Change and Industrial Development in Western Europe from 1750 to the Present* (Cambridge: Cambridge University Press, 1969), p. 41; J. Mokyr, *The Lever of Riches: Technological Creativity and Economic Progress* (Oxford: Oxford University Press, 1990); Bruland, 'Industrialisation and technological change', pp. 117–46; R. C. Allen, *The British Industrial Revolution in Global Perspective* (Cambridge: Cambridge University Press, 2009); J. A. Goldstone, 'Efflorescences and economic growth in world history: Rethinking the "rise of the West" and the Industrial Revolution', *Journal of World History*, 13 (2002), 323–89; N. F. R. Crafts, 'Steam as a general purpose technology: A growth accounting perspective', *Economic Journal*, 114 (2004), 338–51.

5 J. de Vries, 'The Industrial Revolution and the industrious revolution', *Journal of Economic History*, 54 (1994), 262, and *The Industrious Revolution: Consumer Behaviour and the Household Economy, 1650 to the Present* (Cambridge: Cambridge University Press, 2008); C. Muldrew, *Food, Energy and the Creation of Industriousness: Work and Material Culture in Agrarian England, 1550–1780* (Cambridge: Cambridge University Press, 2011); H-J. Voth, *Time and Work in England 1750–1830* (Oxford: Clarendon Press, 2000). See also R. Samuel, 'Workshop of the world: Steam power and hand technology in mid-Victorian Britain', in J. Hoppit and E.A. Wrigley (eds), *The Industrial Revolution in Britain*, vol. 2 (Oxford: Blackwell, 1994), pp. 197–250; M. Berg, *The Age of Manufactures, 1700–1820: Industry, Innovation and Work in Britain* (2nd edn, London: Routledge, 1994), p. 2; J. Humphries, *Childhood and Child Labour in the British Industrial Revolution* (Cambridge: Cambridge University Press, 2010); Earle, *A City Full of People*, ch. 4.

6 Edgerton, *The Shock of the Old*, pp. 9, 33. See also Joyce and Bennett (eds), *Material Powers*; Latour, *Reassembling the Social*; V. DeJohn Anderson, *Creatures of Empire: How Domestic Animals Transformed Early America*

(Oxford: Oxford University Press, 2004); McShane and Tarr, *The Horse in the City*, p. 14.

7 T. S. Reynolds, *Stronger Than a Hundred Men: A History of the Vertical Water Wheel* (Baltimore: Johns Hopkins University Press, 1983).

8 Edward Walford, *Old and New London: A Narrative of its History, its People, and its Places. Volume 6: The Southern Suburbs* (1873), p. 126; John Stow [updated by John Strype], *A Survey of the Cities of London and Westminster: Containing the Original, Antiquity, Increase, Modern Estate and Government of those Cities Written at first in the Year MDXCVIII by John Stow, Citizen and Native of London*, 2 vols (1720), vol. 1, p. 27; J. Graham-Leigh, *London's Water Wars: The Competition for London's Water Supply in the Nineteenth Century* (London: Francis Boutle, 2000), p. 9.

9 P. Hudson, *The Industrial Revolution* (London: Edward Arnold, 1992), p. 101. See also P. Hudson, (ed.), *Regions and Industries: A Perspective on the Industrial Revolution in Britain* (Cambridge, 1989); Reynolds, *Stronger Than a Hundred Men*; H. Beighton, 'A Description of the water-works at London-Bridge, explaining the draught of Tab. I', in Royal Society of London, *Philosophical Transactions*, 37 (1731–32), 5–12; R. L. Hills, *Power from Wind: A History of Windmill Technology* (Cambridge: Cambridge University Press, 1994), p. 234; A Map of the Parish of St Mary Rotherhith, the survey revised and corrected by John Pullen. Engraved by John Harris, in Stow [updated by Strype], *A Survey of London*, Appendix 1, p. 87; John Rocque, *A Plan of the Cities of London and Westminster, and Borough of Southwark, London* (1747), republished as *The A to Z of Georgian London* (London: London Topographical Society, 1982).

10 George Dodd, *The Food of London: A Sketch of the Chief Varieties, Sources of Supply, Probable Quantities, Modes of Arrival, Processes of Manufacture, Suspected Adulteration, and Machinery of Distribution, of the Food for a Community of Two Millions and a Half* (1856), p. 176.

11 'Gyn' probably derives from the Anglo-Norman *gin*, the shortened variant of the Old French *engin*; see 'gin, n.1.', *OED Online*, Oxford University Press, March 2018, www.oed.com/view/Entry/78357 (accessed 14 January 2019).

12 Charles Dibdin, *The High Mettled Racer* (1831), p. 20. See also Donald, *Picturing Animals in Britain*, p. 221.

13 H. Gravelot, *A Perspective View of the Engine, Now Made Use of for Driving the Piles of the New Bridge at Westminster* (etching and engraving, 1748); Charles Labelye, *A Description of Westminster Bridge. To which are Added, an Account of the Methods Made Use of in Laying the Foundations of its Piers. And an Answer to the Chief Objections, that Have Been Made Thereto* (1751), pp. 30–2.

14 City and Southwark Coroners' Inquests, 1795, LL, LMCLIC650080577, LMA, Ms. CLA/041/IQ/02/008.

15 17th and 18th Century Burney Collection Newspapers. Gale Cengage Learning; LMA, MS 11936–7, Sun insurance company policy registers, 1,262 volumes (1710–1863); OBP.

16 Barnett, *London, Hub of the Industrial Revolution*, ch. 5; L. Clarke, *Building Capitalism: Historical Change and the Labour Process in the Production of the Built Environment* (London: Routledge, 1992); J. Ayres, *Building the Georgian City* (New Haven and London: Yale University Press, 1998); P. E. Malcolmson, 'Getting a living in the slums of Victorian Kensington', *London Journal*, 1 (1975), 28–55.

17 William Henry Pyne, *Costume of Great Britain* (1804), Plate 20; TNA, PROB, Wills Proved at Prerogative Court of Canterbury, 11/581/196, Will of Edmund Lydgold, Brick maker (and farmer) of St Leonard Shoreditch, Middlesex (2/9/1721); PROB, 11/1273/124, Will of James Haygarth, Brickmaker of St Pancras, Middlesex (6/4/1796); LMA, MS 11936/390, Sun policy 606250, Owen Clutton, brickmaker and farmer, Walworth (1792); A. Cox, 'A vital component: Stock bricks in Georgian London', *Construction History*, 13 (1997), 57–66.

18 Schwarz, *London in the Age of Industrialisation*, pp. 32–5; P. Earle, *The Making of the English Middle Class: Business, Society and Family Life in London, 1660–1730* (Berkeley and Los Angeles: University of California Press, 1989), p. 22; F. J. Fisher, *London and the English Economy 1500–1700*, ed. P. J. Corfield and N. B. Harte (London: Hambledon Press, 1990), p. 197; M. Palmer and P. Neaverson, *Industry in the Landscape, 1700–1900* (London: Routledge, 1994), p. 113; A. E. Musson, *The Growth of British Industry* (New York: Holmes and Meier, 1978), p. 50; Barnet, *London, Hub of the Industrial Revolution*, p. 67.

19 See also William Henry Pyne, *Tanner's mill* (drawing, undated). The British Museum, 1871,0812.1687.

20 Benjamin Martin, *The General Magazines of Arts and Sciences, Philosophical, Philological, Mathematical, and Mechanical*, vol. 14 (1765), p. 37; Samuel, 'Workshop of the world', p. 225; P. Deane, *The First Industrial Revolution* (2nd edn, Cambridge: Cambridge University Press, 1979), pp. 15–17, 157; Landes, *The Unbound Prometheus*; J. Mokyr, *The Enlightened Economy: Britain and the Industrial Revolution 1700–1850* (London: Penguin, 2011), pp. 347–8.

21 LMA: MS 11936/379, Sun policy 585820, Joseph King, Long Lane, Southwark (13/7/1791); MS 11936/377, Sun policy 582958, John Leachman, Grange Road, Southwark (1791); MS 11936/377, Sun policy 582278, William Halstone, Long Lane, Southwark (1791).

22 *The Repertory of Arts, Manufactures, and Agriculture. Vol. 7, Second Series* (1805), p. 407; B. Woodcroft, *Patents for Invention: Reference Index, 1617–1853* (2nd edn, 1862), patents 2205, 2537 and 2871; *The Monthly Magazine, and British Register, Part 1. 1798. Vol. 5* (1798), p. 374; *The Repertory of Arts and Manufactures. Vol. 10* (1799), pp. 77–80.

23 A. F. M. Willich, *The Domestic Encyclopaedia: Or, a Dictionary of Facts, and Useful Knowledge, Comprehending a Concise View of the Latest Discoveries, Inventions and Improvements, Chiefly Applicable to Rural and Domestic Economy*, 5 vols (Philadelphia, 1803), vol. 1, p. 177; L. F. Ellsworth, *Craft to National Industry in the Nineteenth Century: A Case Study of the Transformation of the New York State Tanning Industry* (New York: Arno Press, 1975), p. 150; Berg, *The Age of Manufactures*, p. 52.

24 William Smart, tanner, The King's Head, Long Lane: LMA, MS 11936/409, Sun policy 675107 (1798); Samuel Brooks, tanner, Grange Road: LMA, MS 11936/423, Sun policy 725260 (1801); George Choumert, tanner, Bermondsey: LMA, MS 11936/257, Sun policy 384352 (1777); LMA, MS 11936/409, Sun policy 673702 (1798); LMA, MS 11936/485, policy 976966/7 (1821); William Smyth, tanner, Bethell Place, Camberwell Road: LMA, MS 11936/473, Sun policy 929094 (1817); Barnett, *London, Hub of the Industrial Revolution*, pp. 67–8.

25 John Farey, *A Treatise on the Steam Engine, Historical, Practical, and Descriptive* (1827), p. 654; M. B. Rowlands, *The West Midlands from AD 1000* (Harlow: Longman, 1987), p. 236. See also Deane, *The First Industrial Revolution*, pp. 126–7.

26 Sir John Soane's Museum, Adam Albums, Vol.11, No.33, Design for the drawing room at Northumberland Street, early 1770s.

27 Pigot & Co., *London and Provincial New Commercial Directory for 1826–27* (1826); Barnett, *London, Hub of the Industrial Revolution*, p. 168.

28 D. Cruickshank and N. Burton, *Life in the Georgian City* (London: Viking, 1990), pp. 181–3; I. C. Bristow, *Interior House-Painting Colours and Technology, 1615–1840* (New Haven and London: Yale University Press, 1996); J. Ayres, *Domestic Interiors: The British Tradition 1500–1850* (New Haven and London: Yale University Press, 2003); C. Saumarez Smith, *Eighteenth-Century Decoration: Design and the Domestic Interior in England* (London: Weidenfeld and Nicolson, 1993); T. Rosoman, *London Wallpapers: Their Manufacture and Use, 1690–1840* (London: English Heritage, 1992).

29 BMDPD, Heal, 89.55, Anon., Trade card of Joseph Emerton, colour man (etching and engraving; verso: a manuscript bill made out to Charles Hayne Esqr., dated 6 July 1744); C. Fox, *The Arts of Industry in the Age of Enlightenment* (New Haven and London: Yale University Press 2009), p. 1.

30 BMDPD, Heal, 68.99, Anon., Trade card for George Farr, grocer, at the Bee-hive and Three Sugar Loaves in Wood Street near Cheapside, London (etching and engraving, *c.* 1753). See also Heal, 68.45, Murray, Trade card of R. Brunsden, tea dealer, grocer and oilman at the Three Golden Sugar Loaves in St James's Street (etching, 1750–60), which depicts a horse mill grinding cocoa nibs.
31 *Common Sense or The Englishman's Journal* (13 March 1742 and 27 March 1742).
32 *Country Journal or The Craftsman* (13 June 1730 and 14 December 1728).
33 Robert Campbell, *The London Tradesman. Being a Compendious View of All the Trades, Professions, Arts, Both Liberal and Mechanic, Now Practiced in the Cities of London and Westminster* (1747), p. 103.
34 Ayres, *Domestic Interiors*, p. 127 and *Building the Georgian City*, p. 212.
35 Peter Barfoot and John Wilkes, *The Universal British Directory of Trade, Commerce, and Manufacture*, 5 vols (1790–98); Pigot & Co., *London and Provincial New Commercial Directory for 1826–27*; Barnett, *London, Hub of the Industrial Revolution*, p. 89; Robert Campbell, *The London Tradesman*, pp. 106–7.
36 M. Allen, *Charles Dickens and the Blacking Factory* (St Leonards: Oxford-Stockley Publications, 2011); N. Page (ed.), *Charles Dickens: Family History*, 5 vols (London: Routledge/Thoemmes Press, 1999), vol. 1, p. 40; Pigot, *London and Provincial New Commercial Directory for 1826–27*; Luke Hebert, *The Engineer's and Mechanic's Encyclopaedia, Comprehending Practical Illustrations of the Machinery and Processes Employed in Every Description of Manufacture of the British Empire*, 2 vols (1836), vol. 2, p. 91; Barnett, *London, Hub of the Industrial Revolution*, p. 93; LMA, MS 11936/509, Sun policy 1039165, Joseph and William Cooper, whiting manufacturers, Millbank (1825).
37 *Daily Advertiser* (19 April 1775 and 10 September 1773); *Gazetteer & New Daily Advertiser* (10 July 1771); *Oracle & Daily Advertiser* (8 July 1799); The Boulton & Watt Archive and the Matthew Boulton Papers (microfilm), Part 3, Reel 33, Portfolio 5, an undated drawing of Felix Calvert's brewery notes that the horse-wheel was 20 feet 10 inches in diameter.
38 For dyers, see Robert Campbell, *The London Tradesman*, pp. 262–3; LMA, MS 11936/409, Sun policy 670697, Barchard & Hilton, dyers, Southwark (1797); for pasteboard manufacturers, see *Whitehall Evening Post* (14–16 August 1783); *Morning Post & Daily Advertiser* (10 September 1785); A. H. Shorter, *Paper Making in the British Isles: An Historical and Geographical Study* (Newton Abbot: David & Charles, 1971), pp. 38, 112, and *Paper Mills and Paper Makers in England 1495–1800* (Hilversum, Holland: The Paper Publications Society, 1957), ch. 4.

39 City and Southwark Coroners' Inquests, 1795, LL, LMCLIC650080577; John Culme, *Nineteenth-Century Silver* (London: Hamlyn Publishing Group, 1977), pp. 14–15.

40 George Gregory, *A New and Complete Dictionary of Arts & Sciences*, 3 vols (New York, 1819), vol. 2, 'Ordnance'; A. Saint and P. Guillery, *Survey of London. Volume 48: Woolwich* (New Haven and London: Yale University Press, 2012), pp. 14–19; O. F. G. Hogg, *The Royal Arsenal: Its Background, Origin, and Subsequent History*, 2 vols (London: Oxford University Press, 1963), vol. 1; M. H. Jackson and C. de Beer, *Eighteenth Century Gunfounding: The Verbruggens at the Royal Brass Foundry, a Chapter in the History of Technology* (Newton Abbot: David & Charles, 1973), pp. 46–51.

41 Jenner, 'From conduit community to commercial network?', pp. 250–72; R. Ward, *London's New River* (London: Historical Publications, 2003), pp. 49–55.

42 Jenner, 'From conduit community to commercial network?', p. 257; TNA, C 5/240/16, Court of Chancery, Six Clerks Office, pleadings, Thomas Foxley and another (plaintiffs) v John Read and others (defendants), 1707; William Maitland, *The History and Survey of London from its Foundation to the Present Time* (1st edn, 1739), p. 622 and 2 vols (2nd edn, 1756), vol. 2, pp. 1264–70.

43 J. W. Gough, *Sir Hugh Myddleton: Entrepreneur and Engineer* (Oxford: Clarendon Press, 1964), ch. 3; G. C. Berry, 'Sir Hugh Myddleton and the New River', in D. Smith (ed.), *Water Supply and Public Health Engineering* (Aldershot: Ashgate, 1999), pp. 46–78; William Ellis, *The Second Part of the Timber-Tree Improved* (1742), pp. 67, 182–3.

44 LMA, ACC/2558/NR1/1–2, New River Company records, Minute books of weekly meetings and general courts 'A', 1769–78 and 1778–86.

45 Matthews, William, *Hydraulica, an Historical and Descriptive Account of the Water Works of London: and the Contrivances for Supplying Other Great Cities, in Different Ages and Countries* (1835), p. 68; *The Encyclopaedia Britannica, or, A Dictionary of Arts, Sciences and General Literature*, 21 vols (7th edn, Edinburgh, 1842), vol. 2, p. 365.

46 John Williams, *An Historical Account of Sub-ways in the British Metropolis, for the Flow of Pure Water and Gas into the Houses of the Inhabitants, Without Disturbing the Pavements* (1828), p. 354. For the most detailed recent survey of the trade's development, see L. Tomory, *The History of the London Water Industry, 1580–1820* (Baltimore: Johns Hopkins University Press, 2017).

47 Maitland, *The History and Survey of London* (1st edn, 1739), p. 395 and 2 vols (2nd edn, 1756), vol. 2, p. 1269. See also George Gregory, *A New and Complete Dictionary of Arts & Sciences*, 3 vols (New York, 1819), vol. 1, 'Boring of water

pipes' and vol. 3 'Pipe-Boring'; Thomas Curtis (ed.), *The London Encyclopaedia or Universal Dictionary of Science, Art, Literature and Practical Mechanics*, 22 vols (1829), vol. 17, p. 421; *Encyclopaedia Britannica* (1842), vol.2, pp. 365–6.

48 LMA, ACC/2558/NR1/1, Minutes of meetings on 17 May 1770, 9 August 1770, 21 August 1770 and 25 June 1772; ACC/2558/NR1/2, Minutes of meetings on 12 May 1785 and 7 July 1785; Graham-Leigh, *London's Water Wars*, p. 17.

49 The New River Company directors ordered the standings of the 'large stable in the Pipe yard' to be enlarged in November 1784 and a month later, a second 'two Stall Stable' was to be built. See LMA, ACC/2558/NR1/1, Minutes of meetings on 31 July 1777; ACC/2558/NR1/2, Minutes of meetings on 17 October 1782, 26 June 1783, 4 November 1784, 23 December 1784 and 24 February 1785; ACC/2558/MW/C/15/341/010, Letter from Richard Cheffins to the board of directors, 3 April 1806.

50 Farey, *A Treatise on the Steam Engine*, p. 654.

51 By 1748, the twelve largest firms accounted for 42 per cent of London's production; this rose to 55 per cent by 1776, to 78 per cent by 1815 and to 85 per cent by 1830, according to P. Mathias, *The Brewing Industry in England 1700–1830* (Cambridge: Cambridge University Press, 1959), pp. 22–7, 551.

52 L. Pearson, *British Breweries: An Architectural History* (London: Hambledon, 1999), p. 27; Mathias, *The Brewing Industry*, pp. 79, 82, 85.

53 William Ellis, *The London and Country Brewer* (5th edn, 1744), pp. 180–1; G. Watkins, *The Complete English Brewer; or, the Whole Art and Mystery of Brewing, in all its Various Branches* (3rd edn, 1770), p. 100.

54 Denis Diderot and Jean Le Rond D'Alembert (eds), *Encyclopédie ou Dictionnaire Raisonné des Arts et des Métiers*, 28 vols (Paris, 1751–1772), vol. 2, pp. 400–6.

55 Mathias, *The Brewing Industry*, pp. 80–2.

56 The Boulton & Watt Archive and the Matthew Boulton Papers (microfilm), Part 3, Reel 33, Portfolio 4, undated drawing of Samuel Whitbread's brewery; Reel 33, Portfolio 22, two drawings of John Calvert's brewery, 1787; Reel 34, Portfolio 10, 'Reverse Copy of the Plan of the Boiler & Cylinder tops &c' for Barclay Perkins brewery, 2 June 1786. See also Mathias, *The Brewing Industry*, pp. 81–2; Fox, *The Arts of Industry*, pp. 104–9.

57 Joseph Delafield, letter to his brother, 1 March 1786, quoted in Mathias, *The Brewing Industry*, p. 93. Mathias cites 'Whitbread Records (Brewery)' but I have been unable to locate this document in LMA, 4453, Whitbread and Company Limited brewery records (1742–2000).

58 The Boulton & Watt Archive and the Matthew Boulton Papers (microfilm), Part 13: Boulton & Watt Correspondence and Papers, Reel 239/3/375,

General Correspondence, G, item 13, Henry Goodwyn (London) letter to Boulton & Watt (Birmingham), 7 July 1784; also quoted in Mathias, *The Brewing Industry*, p. 89.

59 LMA, B/THB/B/003–005 and 006–025/A, Rest books of the Black Eagle brewery, Spitalfields, 1759–80 and 1790–1836.

60 Boulton and Watt's engineering drawings (*c.* 1775–1802) suggest that ten London breweries were operating Sun and Planet steam engines by the end of 1799 and John Farey recorded that seventeen breweries had adopted steam power by 1805; see the Boulton & Watt Archive and the Matthew Boulton Papers (microfilm), Part 3, Reel 33, and Farey, *A Treatise on the Steam Engine*, p. 654.

2 Draught horse

1 D. H. Aldcroft and M. J. Freeman (eds), *Transport in the Industrial Revolution* (Manchester: Manchester University Press, 1983); T. Barker and D. Gerhold, *The Rise and Rise of Road Transport, 1700–1990* (Cambridge: Cambridge University Press, 1995); W. Albert, *The Turnpike Road System in England 1663–1840* (Cambridge: Cambridge University Press, 1972); D. Bogart, 'Turnpike trusts and the transportation revolution in 18th century England', *Explorations in Economic History*, 42 (2005), 479–508.

2 Joseph Nightingale, *London and Middlesex: or, an Historical, Commercial, & Descriptive Survey of the Metropolis of Great-Britain: Including Sketches of its Environs, and a Topographical Account of the Most Remarkable Places in the Above County… Vol. 3* (1815), p. 476.

3 Francis Place, diary entry for 22 November 1827, in British Library, Add. MS 27828: 'Place Papers', vol. 40, Manners and Morals, 4, fol. 8.

4 29 Geo. II, c. 88 (1756); *A Description of England and Wales… Vol. 6* (1769), p. 82.

5 *Lloyd's Evening Post* (17–20 March 1758).

6 T. May, *Gondolas and Growlers: The History of the London Horse Cab* (Stroud: Alan Sutton, 1995); M. S. R. Jenner, 'Circulation and disorder: London streets and hackney coaches, *c.* 1640–*c.* 1740', in T. Hitchcock and H. Shore (eds), *The Streets of London: From the Great Fire to the Great Stink* (London: Rivers Oram Press, 2003), pp. 40–58; T. C. Barker and M. Robbins, *A History of London Transport: Passenger Travel and the Development of the Metropolis: Vol. 1: The Nineteenth Century* (London: Allen & Unwin, 1963), p. 4; Aldcroft and Freeman (eds), *Transport in the Industrial Revolution*, p. 54.

7 D. Gerhold, *Road Transport Before the Railways: Russell's London Flying Waggons* (Cambridge: Cambridge University Press, 1993), p. 1; J. A.

Chartres, 'Road carrying in England in the seventeenth century: Myth and reality', *EcHR*, 2nd series, 30 (1977), 73–94; J. A. Chartres and G. L. Turnbull, 'Road transport', in Aldcroft and Freeman (eds), *Transport in the Industrial Revolution*, pp. 64–99; Barker and Gerhold, *The Rise and Rise of Road Transport*; C. Wilson, *England's Apprenticeship, 1603–1763* (London: Longman, 1965); H. J. Dyos and D. H. Aldcroft, *British Transport: An Economic Survey from the Seventeenth Century to the Twentieth* (Leicester: Leicester University Press, 1969); J. A. Chartres, 'The eighteenth-century English inn: A transient "Golden Age"?', in B. Kümin and B. A. Tlusty (eds), *The World of the Tavern: Public Houses in Early Modern Europe* (Aldershot: Ashgate, 2002), pp. 205–26; Mokyr, *The Enlightened Economy*, pp. 198–219; Barnett, *London, Hub of the Industrial Revolution*, p. 183.

8 John Gay, *Trivia: Or, the Art of Walking the Streets of London* (1716); Pierre Jean Grosley, *A Tour to London: Or, New Observations on England, and its Inhabitants. By M. Grosley, F.R.S. Member of the Royal Academies of Inscriptions and Belles Lettres* (trans.), Thomas Nugent, 3 vols (Dublin, 1772), vol. 1, pp. 105–6; P. J. Corfield, 'Walking the city streets: The urban odyssey in eighteenth-century England', *Journal of Urban History*, 16 (1990), 132–74.

9 George Culley, *Observations on Live Stock, Containing Hints for Choosing and Improving the Best Breeds of the Most Useful Kinds of Domestic Animals* (1786), pp. 11–12, 18. See also N. Russell, *Like Engend'ring Like: Heredity and Animal Breeding in Early Modern England* (Cambridge: Cambridge University Press, 2006), pp. 110–11.

10 William Marshall, *The Rural Economy of the Midland Counties; Including the Management of Livestock in Leicestershire and its Environs*, 2 vols (1790), vol. 1, pp. 306–13. See also Arthur Young, *The Farmer's Tour Through the East of England. Being the Register of a Journey Through Various Counties of this Kingdom, to Enquire into the State of Agriculture, &c.*, 4 vols (1771), vol. 1, pp. 119–20; William Youatt, *The Horse; With a Treatise on Draught and a Copious Index* (1831), pp. 98–102.

11 D. Gerhold, 'The growth of the London carrying trade, 1681–1838', *EcHR*, 2nd series 41 (1988), 392–410; Chartres, 'Road carrying in England', pp. 73–94; Chartres and Turnbull, 'Road transport', pp. 64–99; Gerhold, *Road Transport Before the Railways*.

12 30 Geo. II, c. 22 (1757).

13 E. Bennett, *The Worshipful Company of Carmen of London* (3rd edn, Buckingham: Barracuda Books for the Company, 1982).

14 *Gazetteer & New Daily Advertiser* (27 February 1772).

15 Clarke, *Building Capitalism*.

16 *The Universal British Directory of Trade, Commerce, and Manufacture*; Richard Horwood, *Plan of the Cities of London and Westminster, with the Borough of Southwark and parts adjoining showing every house* (3rd edn, 1813), republished as *The A to Z of Regency London* (London: London Topographical Society, 1985).

17 OBP, September 1809, Benjamin Hall (t18090920–159).

18 *London Daily Advertiser* (6 February 1752); *General Advertiser* (25 September 1752); *Daily Advertiser* (11 March 1772).

19 *St James's Chronicle* (30 April 1761).

20 LMA, CLA/006/AD/04/004, Proceedings, Commissioners of Sewers and Pavements of the City of London, 3 February 1767; Clarke, *Building Capitalism*, p. 199.

21 Stove-grate-maker: *Oracle & Daily Advertiser* (10 April 1799); tallow-chandler: *World* (22 March 1787); grocers: *Gazetteer & New Daily Advertiser* (16 October 1775) and *World* (13 April 1790); OBP, February 1809, John Williams (t18090215–76) and September 1809, Mary Tyler (t18090920–6); fellmongers: *Morning Post & Gazetteer* (26 November 1800); stationers: *World* (2 February 1789); flour miller: OBP, June 1809, George Tylor, John Cross Blain (t18090626–68); baker: OBP, April 1814, Samuel Judah, David Davis (t18140420–20); butchers: OBP, June 1783, William Jenkins (t17830604–3); June 1788, Thomas Salmons (t17880625–12); September 1810, John Warren (t18100919–43); silk dyer: OBP, January 1796, John Graham, George Hooker, James Clements (t17960113–41); gardeners: OBP, May 1786, John Delove (t17860531–2); *Public Advertiser* (2 July 1755); undertakers: *Morning Post & Daily Advertiser* (4 December 1784) and *Oracle & Daily Advertiser* (17 July 1799).

22 OBP, September 1795, John Yarmouth (t17950916–98).

23 OBP: October 1765, Robert Holmes (t17651016–15); September 1795, Thomas Pragnall (t17950916–57); September 1769, Joseph Blewmore (t17690906–107); October 1799, William Holyoak (t17991030–67).

24 Henry Mayhew, *London Labour and the London Poor. Vol. 1. The London Street-Folk* (1851), pp. 28–30; S. Shesgreen, *Images of the Outcast: The Urban Poor in the Cries of London* (Manchester: Manchester University Press, 2002), p. 129.

25 OBP: May 1766, Andrew Welch (t17660514–40); February 1785, Robert Roberts, William Blann (t17850223–22); June 1796, Richard Ludman, Ann Rhodes, Eleanor Hughes, Mary Baker (t17960622–8); February 1797, Thomas Gentleman, William Freeman (t17970215–37); January 1818, Richard Allen, John Nightingale (t18180114–10).

26 Bodleian Library, University of Oxford, John Johnson Collection, Trade Cards 5 (2), William Lewis, 'Tallow-Chandler at the Sun' (1754); BMDPD, Heal, 36.38, Anon., Draft trade card of Robert Stone, nightman at the Golden Pole the Upper end of Golden Lane Near Old Street (1745); BMDPD, Heal, 36.36: Anon., Draft trade card of Robert Stone, nightman and rubbish carter (1751).

27 E. H. Dixon, *King's Cross, London: The Great Dust-Heap, next to Battle Bridge and the Smallpox Hospital* (watercolour, 1837), Wellcome Library, London.

28 B. Maidment, *Dusty Bob: A Cultural History of Dustmen, 1780–1870* (Manchester: Manchester University Press, 2007), p. 23.

29 OBP, May 1787, Daniel Hands (t17870523–21); [Thomas Legg], *Low-Life; or One Half of the World Knows Not How the Other Half Live... In the Twenty-four Hours, Between Saturday-Night and Monday-Morning. In a True Description of a Sunday, as it is Usually Spent Within the Bills of Mortality* (3rd edn, 1764), p. 99.

30 *Gazetteer & New Daily Advertiser* (14 May 1774); OBP, January 1780, Emanuel Isaacs (t17800112–16).

31 The Old Bailey Proceedings also record references to errand carts for Islington, Ponders End, Peckham, Woolwich, Lambeth and Vauxhall, Clapham, Bow, Holloway, Highgate, and Stanmore. See OBP: April 1782, Thomas Lewington (t17820410–39); September 1785, Thomas Lewington (t17850914–114); December 1786, Thomas Levington (t17861213–59); February 1800, Elizabeth Crouch (t18000219–78); February 1803, James Nowland, John Price (t18030216–38); December 1806, John Spencer (t18061203–55); June 1808, David Swinton (t18080601–25); June 1808, Thomas Hutchins (t18080601–76); February 1810, James Williams (t18100221–63); May 1811, John George (t18110529–29); January 1813, James Smith (t18130113–35); September 1816, John Smith (t18160918–70); December 1817, John Jones, Peter Davis (t18171203–96); January 1819, John Williams (t18190113–82); January 1819, Edward Mitchell (t18190113–131). For the trials involving tobacco and indigo, see OBP, December 1800, Ann Metcalf (t18001203–66); January 1808, John Turner (t18080113–45).

32 Bennett, *The Worshipful Company of Carmen*, p. 39; Great Britain Coal Commission, *Report of the Commissioners Appointed to Inquire into the Several Matters Relating to Coal in the United Kingdom, Vol.1: General Report and Twenty-two Sub-reports* (1871), appendix 15; OBP, September 1805, Charles Deakin (t18050918–121) and February 1849, George Williams (t18490226–819). On the coal trade in London, see T. S. Ashton and J. Sykes, *The Coal*

Industry in the Eighteenth Century (Manchester: Manchester University Press, 1929), pp. 202–10; *The Universal British Directory of Trade, Commerce, and Manufacture*; *Daily Advertiser* (9 December 1772); *Gazetteer & New Daily Advertiser* (21 February 1780); Henry Mayhew, 'Letter XXI', in *The Morning Chronicle* (28 December 1849).

33 OBP, September 1805, Charles Deakin (t18050918–121).

34 LL, WACWIC652110315 and Westminster Abbey Muniment Room, Westminster Coroners' Inquest into the death of Heller Thoume, 1771; BMDPD, Banks, 44.25, Anon., Draft trade card of J. Williams Junior, coal merchant, No. 2 Beaumont's Buildings, St. George's East, and at Bush Wharf, Southwark (*c.* 1790); W. J. Gordon, *The Horse World of London* (1893), pp. 130–2.

35 Ashton and Sykes, *The Coal Industry*, p. 62.

36 Mayhew, 'Letter XXI'.

37 BMDPD, Banks, 44.11, Anon., Draft trade card of Benjamin Levy, coal merchant (engraving, *c.* 1780). Arguably the most impressive depiction of London's coal horses is, however, Theodore Gericault's *The Coal Wagon, or Le Chariot, Route de Londres* (drawing and watercolour, 1820–21). The British Museum.

38 A. Velkar, *Markets and Measurements in Nineteenth-Century Britain* (Cambridge: Cambridge University Press, 2012), p. 100; Gordon, *The Horse World of London*, pp. 129–30; Mayhew, 'Letter XXI'. See also F. M. L. Thompson, *Victorian England: The Horse-Drawn Society: An Inaugural Lecture* (London: Bedford College, University of London, 1970), p. 13.

39 Mathias, *The Brewing Industry*, pp. 22, 79, 551.

40 George Stubbs, *Horse Frightened by a Lion* (oil on canvas, exhibited 1763), Tate; S. Deuchar, *Paintings, Politics and Porter: Samuel Whitbread II and British Art* (exhibition catalogue, London: Museum of London, 1984), pp. 46–7, 77.

41 Dean Wolstenholme the elder, *Red Lion Brewhouse, East Smithfield* (mezzotint after oil on canvas, 1805) and *A Correct View of the Golden Lane Genuine Brewery* (mezzotint after oil on canvas, 1807); Dean Wolstenholme the younger, *A correct view of the Hour Glass Brewery belonging to Messrs. Calvert and Co* (coloured aquatint after oil on canvas, 1821) and *Messrs. Truman, Hanbury and Buxton's Black Eagle Brewery* (oil on canvas, exhibited at the Royal Academy in 1822). On both artists, see W. Gilbey, *Animal Painters of England from the Year 1650: A Brief History of Their Lives and Works*, 2 vols (London: Vinton & Co., 1900), vol. 2, pp. 244–59.

42 John Nost Sartorius, *Old Brown aged 35, a Dray-Horse in a Brewery Yard* (oil on canvas, 1798), private collection; reproduced in E. Moncrieff, S. Joseph

and I. Joseph, *Farm Animal Portraits* (Woodbridge: Antique Collectors' Club, 1996).

43 James Egan after John Christian Zeitter, *Pirate and Outlaw* (etching and engraving, *c.* 1818); Gilbey, *Animal Painters*, vol. 1, pp. 176–89.

44 Youatt, *The Horse*, pp. 100–1. See also John Middleton, *View of the Agriculture of Middlesex; with Observations on the Means of its Improvement, and Several Essays on Agriculture in General. Drawn up for the Consideration of the Board of Agriculture* (1798), p. 360; Mathias, *The Brewing Industry*, pp. 78–9.

45 Wordsworth, *The Prelude*, p. 109.

46 P. D. McGreevy and A. N. McLean, 'Behavioural problems with the ridden horse', in D. S. Mills and S. M. McDonnell (eds), *The Domestic Horse: The Origins, Development and Management of its Behaviour* (Cambridge: Cambridge University Press, 2005), p. 198; see also G. H. Waring, 'Agonistic behaviour', in G. H. Waring (ed.), *Horse Behaviour* (Norwich, NY: Noyes Publications, 2003), pp. 253–69.

47 *Lloyd's Evening Post* (17 February 1762).

48 *London Chronicle* (10 July 1764).

49 1 Geo. II, Stat. 2, c. 52 (1715); 24 Geo. II, c. 43 (1750).

50 *Old England* (16 July 1748); *London Daily Advertiser* (15 June 1751); 24 Geo. II, c. 43 (1750); R. Paulson, *The Life of Henry Fielding: A Critical Biography* (Oxford: Blackwell, 2000), pp. 271, 276.

51 Young, *The Farmer's Tour*, vol. 1, p. 120.

52 *The Sporting Magazine*, vol. 7, 3rd series (1846), p. 411.

53 A. Wolff and M. Hausberger, 'Learning and memorisation of two different tasks in horses: The effects of age, sex and sire', *Applied Animal Behaviour Science*, 46 (1996), 137–43; S. L. Marinier and A. J. Alexander, 'The use of a maze in testing learning and memory in horses', *Applied Animal Behaviour Science*, 39 (1994), 177–82; Greene, *Horses at Work*, p. 22.

54 *The Gentleman's Magazine*, 30 (1760), pp. 527–9; Mathias, *The Brewing Industry*, pp. 105, 117–23.

55 LMA, 4453/B/12/002, Rest book of Whitbread's brewery, Chiswell Street, 1800; OBP, June 1764, James Mannen (t17640607–49). See also, John Allen, diary entries, 4 and 18 June 1777, in C. Y. Sturge (ed.), *Leaves from the Past: The Diary of John Allen, Sometime Brewer of Wapping 1757–1808* (Bristol: Arrowsmith, 1905), pp. 63–64, 78; Louis Simond, *Journal of a Tour and Residence in Great Britain, During the Years 1810 and 1811*, 2 vols (Edinburgh, 1817), vol. 1, pp. 182–4; T. R. Gourvish and R. G. Wilson, *The British Brewing Industry, 1830–1980* (Cambridge: Cambridge University Press, 1994), p. 142.

56 The Black Eagle brewery rest books record the number of dray and mill horses in the firm's service after 1775; a stable book for Barclay Perkins

brewery provides a weekly stock-take of the company's dray horses from September 1827 to April 1839. See LMA, B/THB/B/001–025/A, Rest books of the Black Eagle brewery, Spitalfields, 1741–1836 and ACC/2305/1/1300, Stable book of Barclay Perkins brewery, Southwark, September 1827–April 1839; Voth, *Time and Work in England*, p. 122.

57 LMA, ACC/2305/01/0834, A plan of an estate belonging to Henry Thrale esq. situate in Park Street, Maid Lane and Castle Lane, surveyed anno 1774 by George Gwilt, copy, photographic negative, 1983.

58 LMA, ACC/2305/01/0159/001, Rest book, Anchor brewery, 1780; Pearson, *British Breweries*, p. 34.

59 LMA, 4453/B/12/002; B. Spiller, 'The Georgian brewery', *Architectural Review*, 122 (1957), 321; see also BMDPD, 1880, 1113.4854, *Plan of the Brewhouse in Liquor-Pond Street*, hand-coloured engraving, James Basire II, 1796, which details 65 horse stalls arranged in two stable blocks.

60 *The Civil Engineer and Architect's Journal*, 1:1 (1837), pp. 47–8.

61 H. Maxwell (ed.), *The Creevey Papers: A Selection From the Correspondence and Diaries of the Late Thomas Creevey, MP*, 2 vols (Cambridge: Cambridge University Press, 2012), vol. 2, p. 71.

62 *The Civil Engineer and Architect's Journal*, pp. 48–50.

63 Simond, *Journal of a Tour*, vol. 1, pp. 183–4.

64 P. Warde, *Energy Consumption in England and Wales, 1560–2000* (Naples: Instituto di Studi sulle Società del Mediterraneo, 2007), p. 44; A. E. Collins, 'Power availability and agricultural productivity in England and Wales, 1840–1939', in B. J. P. van Bavel and E. Thoen (eds), *Land Productivity and Agro-Systems in the North Sea Area* (Turnhout: Brepols, 1999), pp. 216–17.

65 N. E. Robinson (ed.), *Current Therapy in Equine Medicine*, 4 vols (4th edn, Philadelphia: W. B. Saunders, 1997), p. 202; H. M. Clayton, *Conditioning Sport Horses* (Saskatoon: Sport Horse Publications, 1991), p. 158.

66 The Black Eagle brewery's rest books for 1781–89 do not survive and the rest book for 1790 does not provide a breakdown of the foods given; LMA, B/THB/B/004–014, Rest books, 1767–1817.

67 Middleton, *View of the Agriculture of Middlesex*, pp. 362, 366; Youatt, *The Horse*, p. 353. See also Gerhold, *Road Transport Before the Railways*, p. 130.

68 J. Thirsk (ed.), *The Agrarian History of England and Wales, vol. 5, 1640–1750: 2. Agrarian Change* (Cambridge: Cambridge University Press, 1985), Table 13.4; B. A. Holderness, 'Prices, productivity, and output', in G. E. Mingay (ed.), *The Agrarian History of England and Wales, vol. 6: 1750–1850* (Cambridge: Cambridge University Press, 1989), pp. 92–109, 124–5, Table I.5;

Peter Foot, *General View of the Agriculture of the County of Middlesex, with Observations on the Means of their Improvement* (1794), p. 58.

69 *Morning Post & Fashionable World* (26 March 1795).

70 LMA, ACC/2305/01/0176/004, List of workers, 1791–97, a note in rest book of Barclay Perkins brewery, Southwark, 1797.

71 Quoted in Middleton, *View of the Agriculture of Middlesex*, pp. 364–5.

72 LMA, B/THB/B/014–015, Rest books, 1814–18; James Clark, *Observations Upon the Shoeing of Horses: Together with a New Inquiry into the Causes of Diseases in the Feet of Horses*, 2 vols (3rd edn, Edinburgh, 1782), vol. 1, p. 62.

73 LMA, ACC/2305/1/1300, Stable book of Barclay Perkins brewery; Simond, *Journal of a Tour*, vol. 1, pp. 182–4.

74 LMA, ACC/2305/01/0834, A plan of an estate belonging to Henry Thrale.

75 Spiller, 'The Georgian brewery', 321; Mathias, *The Brewing Industry*, p. 79; *The Penny Magazine of the Society for the Diffusion of Useful Knowledge. New Series*, 10 (1841), p. 128.

76 Gordon, *The Horse World of London*, p. 86; Middleton, *View of the Agriculture of Middlesex*, p. 367.

77 John Lawrence, *A Philosophical and Practical Treatise on Horses, and on the Moral Duties of Man Towards the Brute Creation*, 2 vols (1796–98), vol. 1, pp. 306–10.

78 This debate began with E. J. Hobsbawm, 'The British standard of living, 1790–1850', *EcHR*, 10 (1957), 46–68; R. M. Hartwell, 'The rising standard of living in England, 1800–1850', *EcHR*, 13 (1961), 397–416; E. J. Hobsbawm, 'The standard of living during the Industrial Revolution: A discussion', *EcHR*, 16 (1963), 119–34; R. M. Hartwell, 'The standard of living: An answer to the pessimists', *EcHR*, 16 (1963), 135–46. Subsequent contributions have included P. H. Lindert and J. G. Williamson, 'English workers' living standards during the Industrial Revolution: A new look', *EcHR*, 36 (1983), 1–25; S. Horrell and J. Humphries, 'Old questions, new data, and alternative perspectives: Families' living standards in the Industrial Revolution', *Journal of Economic History*, 52 (1992), 849–80; C. H. Feinstein, 'Pessimism perpetuated: Real wages and the standard of living in Britain during and after the Industrial Revolution', *Journal of Economic History*, 58 (1998), 625–58; Voth, *Time and Work in England*; de Vries, *The Industrious Revolution*.

79 Jenner, 'Circulation and disorder'.

80 *Public Advertiser* (11 September 1767).

81 OBP, May 1736, John Maccoon (t17360505–61).

82 1 Geo. II, Stat. 2, c. 52 (1715); 24 Geo. II, c. 43 (1750); 30 Geo. II, c. 22 (1757).

83 LMA, B/THB/B/012 and 025/A, Rest books, 1809–13 and 1835–36; ACC/2305/1/1300, Stable book of Barclay Perkins brewery.

84 Gourvish and Wilson, *The British Brewing Industry*, pp. 141, 144.

85 *London Chronicle* (26 May 1787).

86 'A steam engine, which equals the force of twenty-eight horses' and 'fifty-eight magnificent horses, each worth £50 ... [which] are employed to carry the liquor about London and its surroundings'. Author's translation of Marc-Auguste Pictet, *Voyage de Trois Mois, en Engleterre, en Écosse et en Irelande* (Geneva, 1802), pp. 305–6.

87 Truman Hanbury Buxton and Company, *Trumans the Brewers, 1666–1966* (London: Newman Neame, 1966), p. 35.

3 Animal husbandry

1 *Diary or Woodfall's Register* (18 September 1790).

2 [Thomas Cox], *Magna Britannia Antiqua & Nova: Or, a New, Exact, and Comprehensive Survey of the Ancient and Present State of Great-Britain*, 6 vols (1738), vol. 3, p. 1; W. K. Jordan, *The Charities of London 1480–1660* (London: Allen & Unwin, 1960), p. 46.

3 C. Phythian-Adams, 'Milk and soot: The changing vocabulary of a popular ritual in Stuart and Hanoverian London', in D. Fraser and A. Sutcliffe (eds), *The Pursuit of Urban History* (London: Edward Arnold, 1983), pp. 83–104; P. Guillery, *The Small House in Eighteenth-Century London: A Social and Architectural History* (New Haven and London: Yale University Press, 2004).

4 Mingay (ed.), *The Agrarian History of England and Wales, vol. 6*; R. A. C. Parker, *Coke of Norfolk: A Financial and Agricultural Study 1707–1842* (Oxford: Clarendon Press, 1975); R. Trow-Smith, *A History of British Livestock Husbandry, 1700–1900* (London: Routledge & Kegan Paul, 1959); G. E. Fussell, 'Science and practice in eighteenth-century British agriculture', *Agricultural History*, 43 (1969), 7–18; D. Brown, 'Reassessing the influence of the aristocratic improver: the example of the fifth Duke of Bedford (1765–1802)', *AgHR*, 47 (1999), 182–95; D. L. Wykes, 'Robert Bakewell (1725–1795) of Dishley: Farmer and livestock improver', *AgHR*, 52 (2004), 38–55.

5 Cockayne, *Hubbub*, pp. 147–8, 192–3, 213–15; D. Gray, *London's Shadows: The Dark Side of the Victorian City* (London: Continuum, 2010), p. 76; T. Hunt, *Building Jerusalem: The Rise and Fall of the Victorian City* (London: Weidenfeld & Nicolson, 2004), pp. xxix–xxx; S. Wise, *The Blackest Streets: The Life and Death of a Victorian Slum* (London: Vintage Books, 2009); C. Hamlin, *Public Health and Social Justice in the Age of Chadwick: Britain, 1800–1854* (Cambridge: Cambridge University Press, 1998).

6 Thomas, *Man and the Natural World*; Trow-Smith, *A History of British Livestock Husbandry*, p. 18.

7 P. J. Atkins, 'The milk trade of London, *c.* 1790–1914' (PhD thesis, University of Cambridge, 1977), pp. 1, 6; E. J. T. Collins, 'Introduction', in E. J. T. Collins (ed.), *The Agrarian History of England and Wales, vol. 7: 1850–1914 (Part 1)* (Cambridge: Cambridge University Press, 2000), p. 21.

8 Berger, 'Why look at animals?', p. 16, emphasis in original.

9 D. Valenze, *Milk: A Local and Global History* (New Haven and London: Yale University Press, 2011), pp. 85–99.

10 *The Westminster Magazine*, vol. 6 (March 1778), pp. 64–6.

11 Foot, *General View of the Agriculture of the County of Middlesex*, p. 80; Thomas Baird in Arthur Young (ed.), *Annals of Agriculture and Other Useful Arts* (1793), vol. 21, p. 117.

12 *Gazetteer* (26 January 1773); *London Daily Post* (25 March 1736). See also *London Evening Post* (18–20 February 1735).

13 LMA, MS/11937–7; Burney Collection; LL.

14 For Samuel Harvey, see *Post Boy* (21–24 June 1712) and *Read's Weekly Journal or British Gazetteer* (15 December 1750). For Richard Ferryman, see TNA, PROB 11/586/348 and 11/731/111, Wills Proved at Prerogative Court of Canterbury (16 August 1722 and 16 January 1744). For John Unthank, see *London Evening Post* (16–18 January 1750); LMA, MS 11936/368, Sun policy 571787 (1790).

15 H. Steinfeld, P. Gerber, T. Wassenaar, V. Castel, M. Rosales and C. de Haan, *Livestock's Long Shadow: Environmental Issues and Options* (Rome: Food and Agriculture Organization of the United Nations, 2006), p. 33.

16 Atkins, 'The milk trade of London', pp. 24–6 and 'London's intra-urban milk supply, circa 1790–1914', *Transactions of the Institute of British Geographers*, 2 (1977), 383–99.

17 Atkins, 'The milk trade of London', p. 28; L. Martindale, 'Demography and land use in the late seventeenth and eighteenth centuries in Middlesex' (PhD thesis, University of London, 1968), pp. 319–20; Daniel Lysons, *The Environs of London: Being an Historical Account of the Towns, Villages, and Hamlets, within Twelve Miles of that Capital*, 4 vols (1795–96), vol. 3, pp. 123–4; John Nelson, *The History, Topography, and Antiquities of the Parish of St. Mary Islington, in the County of Middlesex* (1811), pp. 106–11.

18 LMA, MS/11937–7; Burney Collection; TNA, PROB; OBP; LL, the most valuable datasets being the Westminster poll books and the Middlesex Sessions Papers.

19 Anon., 'Cows for the supply of London', in Young (ed.), *Annals of Agriculture* (1793), vol. 21, pp. 530; Anon., *A Treatise on Milk, as an Article of the First*

Necessity to the Health and Comfort of the Community: A Review of the Difference Methods of Production; and Suggestions Respecting the Best Means of Improving its Quality, Reducing its Price, and Increasing its Consumption (1825), p. 81; William Youatt, *Cattle; Their Breeds, Management, and Diseases* (1834), p. 264; David Brewster (ed.), *The Edinburgh Encyclopaedia*, 18 vols (Philadelphia, 1832), vol. 13, p. 257; Atkins, 'The milk trade of London', pp. 22–31, 81.

20 Tobias Smollett, *The Expedition of Humphry Clinker*, ed., L. M. Knapp, revised by P.-G. Boucé (1771; Oxford: Oxford University Press, 1998), p. 121.

21 Cockayne, *Hubbub*, pp. 147–8, 192–3, 213–15; C. McNeur, 'The "Swinish Multitude": Controversies over hogs in Antebellum New York City', *Journal of Urban History*, 37 (2011), 639–60 and *Taming Manhattan: Environmental Battles in the Antebellum City* (Cambridge, MA and London: Harvard University Press, 2014).

22 Cockayne, *Hubbub*, p. 193; Trow-Smith, *A History of British Livestock Husbandry*, pp. 18, 217.

23 Stow [updated by Strype], *A Survey of London*, Appendix 1, Ch. 8, p. 49.

24 Stow [updated by Strype], *A Survey of London*, Book 5, p. 450; E. G. Dowdell, *A Hundred Years of Quarter Sessions: The Government of Middlesex From 1660 to 1760* (Cambridge: Cambridge University Press, 1932), p. 123; Maitland, *The History and Survey of London* (1760), vol.1, p. 456.

25 2 Will. & Mar., sess. 2, c. 8 (1690); 8 & 9 Will. III, c. 37 (1696); W. J. Hardy (ed.), *Middlesex County Records: Calendar of the Sessions Books 1689 to 1709* (London: Richard Nicholson, 1905), pp. 27, 28, 33–40, 44–50, 56–61.

26 LMA, MJ/O/C/001, Middlesex Sessions, General orders of court, 6 December 1720 (LL, LMSMGO400000233).

27 S. Webb and B. Webb, *English Local Government from the Revolution to the Municipal Corporations Act: The Parish and the County* (London: Longmans, Green and Co., 1906), p. 531.

28 *Public Advertiser* (6 March 1762 and 6 August 1768).

29 LL, WCCDMV362010176, St Clement Danes vestry minutes, 3 October 1799, Westminster Archives Centre, Ms. B1074.

30 OBP, February 1794, William Sullivan (t17940219–77).

31 James Bell, *A System of Geography, Popular and Scientific, or a Physical, Political, and Statistical Account of the World and its Various Divisions*, 6 vols (Glasgow, 1832), vol. 3, p. 102; John Feltham, *The Picture of London* (1807), p. 37.

32 Cronon, *Nature's Metropolis*, pp. 225–6.

33 J. M. Beattie, *Crime and the Courts in England, 1660–1800* (Oxford: Clarendon Press, 1986), pp. 309–13.

34 OBP, February 1794, William Sullivan (t17940219–77).

35 TNA, PROB 2, Inventories compiled before 1661 (1417–1668) and PROB 4, Engrossed Inventories Exhibited from 1660 (1660–c1720); LMA, CLA/004/02/001–073, Minute books of the Mansion House Justice Room, City of London, 1784–1821 and CLA/005/01/001–055, Minute books of the Guildhall Justice Room, City of London, 1752–1796; C. Rawcliffe, *Urban Bodies: Communal Health in Late Medieval English Towns and Cities* (Woodbridge: The Boydell Press, 2013), p. 157–60; D. Gray, *Crime, Prosecution and Social Relations: The Summary Courts of the City of London in the Late Eighteenth Century* (Basingstoke: Palgrave Macmillan, 2009).

36 TNA, KB 10/33, Court of King's Bench: Crown Side: London and Middlesex Indictment Files, 1759–1761, Box 1, Item 33, case of John Jolly of St Clement Danes, 1760; Box 1, Item 78, case of Joseph Cowling of St John, Hackney, 1761; Box 1, Item 82: case of William Duck of Hampstead, 1761; Box 3, Item 17, case of John Hardy of Spitalfields, 1761; KB 10/47, Court of King's Bench: Crown Side: London and Middlesex Indictment Files, 1790–1791, case of James Welch of St Margaret, Westminster, Hilary, 1791, and case of Frederick Tasman, Islington, Easter, 1791.

37 TNA, KB 10/33, Box 1, Item 33, 1760; TNA, KB 10/47, Easter 1791.

38 George Harris, *The Life of Lord Chancellor Hardwick: With Selections from his Correspondence, Diaries, Speeches, and Judgements*, 3 vols (1847), vol. 1, p. 269; J. Oldham, *The Mansfield Manuscripts and the Growth of English Law in the Eighteenth Century*, 2 vols (Chapel Hill, NC, and London: University of North Carolina Press, 1992), vol. 2, ch. 15; William Blackstone, *Commentaries on the Laws of England*, ed. Wilfrid Prest, 4 vols (Oxford, 1765–69; Oxford: Oxford University Press, 2016), vol. 3, p. 144.

39 Oldham, *The Mansfield Manuscripts*, vol. 2, p. 892.

40 Mayhew, *London Labour and the London Poor*, 3 vols (1861–62), vol. 2, p. 193.

41 Harris, *The Life of Lord Chancellor Hardwick*, vol. 1, p. 269; *Daily Journal* (19 February 1734).

42 LMA, HFCS/018, Holborn and Finsbury Commission of Sewers Minutes, 1763–98.

43 W. C. Baer, 'Housing for the lesser sort in Stuart London: Findings from certificates, and returns of divided houses', *London Journal*, 33 (2008), 64; LMA, HFCS/011, Holborn and Finsbury Commission of Sewers extracts from minutes, copies of warrants, etc., 1683–88, fo. 6; LMA, HFCS/018, Minutes, 18 August 1773, pp. 257–8.

44 LMA, HFCS/018, pp. 107, 222, 257–8; see also LMA, MR/L/SB/001, Middlesex Sessions, Registers of copies of slaughter house licences certified by ministers and churchwardens, 1786–1822.

45 Guillery, *The Small House*, p. 21; C. Spence, *London in the 1690s: A Social Atlas* (London: Centre for Metropolitan History, Institute of Historical Research, University of London, 2000), p. 65; V. Harding, 'Housing and health in early modern London', in V. Berridge and M. Gorsky (eds), *Environment, Health and History* (Basingstoke: Palgrave Macmillan, 2011), p. 28, and 'The population of London, 1550–1700: a review of the published evidence', *London Journal*, 15 (1990), 111–28.

46 Smollett, *The Expedition of Humphry Clinker*, p. 122. See also Cockayne, *Hubbub*, ch. 4; Valenze, *Milk*, p. 137; M. Lane, *Jane Austen and Food* (London: Hambledon Press, 1995), p. 12; K. Colquhoun, *Taste: The Story of Britain Through its Cooking* (London: Bloomsbury, 2008), pp. 193–4.

47 Lysons, *The Environs of London*, vol. 4, pp. 575–76; Richard Parkinson, *Treatise on the Breeding and Management of Live Stock… in which the Principles and Proceedings of the New School of Breeders are Fully and Experimentally Discussed*, 2 vols (1810), vol. 1, pp. 68–9; Anon., 'A day at a London dairy', *The Penny Magazine*, 10 (1841), 301; Nelson, *The History, Topography, and Antiquities of the Parish of St. Mary Islington*, pp. 106–7; Martindale, 'Demography and land use', pp. 72–5; Atkins, 'The milk trade of London', pp. 29–30.

48 P. Mathias, 'Agriculture and the brewing and distilling industries in the eighteenth century', *EcHR*, 5 (1952), 249–50; Mingay (ed.), *The Agrarian History of England and Wales*, vol. 6, p. 255.

49 [Legg], *Low-Life*, p. 90.

50 OBP, May 1803, James Rudd, John Willis (t18030525–38).

51 Youatt, *Cattle*, p. 264; Atkins, 'The milk trade of London', p. 75.

52 *Gazetteer & New Daily Advertiser* (19 February 1782); Richard Bradley, *The Gentleman and Farmer's Guide for the Increase and Improvement of Cattle* (2nd edn, 1732), p. 77; Mingay (ed.), *The Agrarian History of England and Wales*, vol. 6, pp. 169–70, 356–7.

53 Mathias, 'Agriculture and the brewing and distilling industries'; Atkins, 'The milk trade of London', pp. 29, 75.

54 Atkins, 'The milk trade of London', p. 28; Martindale, 'Demography and land use', pp. 319–20; Richard Hodgkinson, journal entry for 16 May 1819, in F. Wood and K. Wood (eds), *A Lancashire Gentleman: The Letters and Journals of Richard Hodgkinson, 1763–1847* (Stroud: Allan Sutton Press, 1992), pp. 186–7.

55 John Rocque, *A Plan of the Cities of London and Westminster, and Borough of Southwark, London* (1747), republished as *The A to Z of Georgian London* (London: London Topographical Society, 1982).

56 LMA, P92/SAV/0444, A survey and valuation of all the lands, buildings, houses, tenements and hereditaments within the parish of Saint Saviour Southwark by John Middleton and Thomas Swithin, surveyors, 1807–8.

57 LMA, MS 11936/478, Sun policy 9622467 (1820); Thomas Allen, *The History and Antiquities of London, Westminster and Southwark, and Parts Adjacent*, 4 vols (1827–29), vol. 4, p. 539.

58 R. Malcolmson and S. Mastoris, *The English Pig: A History* (London and New York: Hambledon and London, 2001), p. 35; Mingay (ed.), *The Agrarian History of England and Wales*, vol. 6, pp. 326–34, 348–50.

59 A. S., *The Husbandman, Farmer and Grasier's Compleat Instructor* (1697), p. 91.

60 Joseph Lucas (trans.), *Kalm's Account of his Visit to England on his way to America in 1748* (1892), p. 411; Mathias, 'Agriculture and the brewing and distilling industries', 254; William James and Jacob Malcolm, *General View of the Agriculture of the County of Surrey* (1794), p. 33.

61 Anon., *An Impartial Enquiry into the Present State of the British Distillery; Plainly Demonstrating the Evil Consequences of Imposing any Additional Duties on British Spirits* (1736), p. 38; Anon., *The Corn Distillery, Stated to the Consideration of the Landed Interest of England* (1783), p. 44.

62 *Journal of the House of Commons, vol. 23: 1737–1741* (1803), pp. 584, 630, and *vol. 24: 1741–45* (1803), pp. 833–6; Mathias, 'Agriculture and the brewing and distilling industries', 251–2; Thomas Wilson, *Distilled Spiritous Liquors the Bane of the Nation: being some considerations humbly offer'd to the legislature* (1736); [Josiah Tucker,] *An Impartial Inquiry into the Benefits and Damages Arising to the Nation from the Present Very Great Use of Low-priced Spiritous Liquors: With Proper Estimates Thereupon, and Some Considerations Humbly Offered for Preventing the Introduction of Foreign Spirits Not Paying the Duties. By J. T. of Bristol* (1751).

63 Lucas (trans.), *Kalm's Account of his Visit to England*, p. 411; see also Bradley, *The Gentleman and Farmer's Guide*, pp. 77–8, and *A General Description of All Trades, Digested in Alphabetical Order* (1747), p. 79.

64 Thomas Pennant, *Of London* (1790), p. 33; James Gilray, *A Pig in a Poke. Whist, Whist'* (hand-coloured etching, 1788), and *Pigs Meat; – or – The Swine Flogg'd Out of the Farm Yard* (hand-coloured etching, 1798).

65 BL, Add MS 39683, [Thomas Cooke], *Observations upon Brewing, Fermentation, and Distillation, with sundry remarks and observations upon erection of corn distillhouses, situation, conveniences, repairs, expences, etc* [apparently unpublished, *c.* 1792?]; Mathias, 'Agriculture and the brewing and distilling industries', 254; James and Malcolm, *General View of the Agriculture of the County of Surrey*, p. 33.

66 Mathias, 'Agriculture and the brewing and distilling industries', 252.

67 TNA, ADM 112/162, Navy Board, Office of Surveyor of Victuals and Victualling Office, Contract Ledger, 1776.

68 D. Morris, *Whitechapel 1600–1800: A Social History of an Early Modern London Inner Suburb* (London: East London History Society, 2011); N. A. M. Rodger, *The Wooden World: An Anatomy of the Georgian Navy* (London: Collins, 1986), ch. 3; P. MacDougall, *London and the Georgian Navy* (Stroud: The History Press, 2013), pp. 102–3.

69 I. Grainger and C. Phillpotts, *The Royal Navy Victualling Yard, East Smithfield, London* (London: Museum of London, 2010), pp. 11, 37, 86, 90; TNA, ADM 110/9, Navy Board, Victualling Office, Out-letters, Letter Book, 1722–27, pp. 73–6.

70 Thomas, *Man and the Natural World*, p. 182.

71 Cobbett, *Rural Rides*, pp. xv, 7, 36–8, 49, 89, 133, 284–8, 296, 322, 355, 377.

72 LMA, WR/PP/1788/1–3/430/43073, Westminster poll books (1788), Parish of St Margaret, Westminster, and St John the Evangelist: John Tice, cow-keeper and shoe maker, Princes Street; *Bell's Weekly Messenger* (14 October 1798); *Morning Chronicle* (18 July 1794).

73 LMA, MS 11936/279, Sun policy 421775, Richard Onion, carman and cowkeeper (1779); LMA, WR/PP/1784/16–20/346/34654, Westminster Poll Books (1784), Parish of St Margaret and St John: William Terry, coachmaster and cowkeeper, St Annes Lane. For victuallers, see *London Gazette* (13–17 May 1755 and 10–14 June 1755); *London Evening Post* (2–5 April 1763); *Gazetteer* (13 May 1769); LMA, MS 11936/268, Sun policy 403811 (1778).

74 LMA, MS 11936/293, Sun policy 445713, William Cornwell, ropemaker and cow-keeper (1781).

75 M. Bowden, G. Brown and N. Smith, *An Archaeology of Town Commons in England: 'A Very Fair Field Indeed'* (Swindon: English Heritage, 2009), pp. 27, 41, 58; H. French, 'Urban common rights, enclosure and the market: Clitheroe Town Moors, 1764–1802', *AgHR*, 51 (2003), 41.

76 I. Darlington (ed.), *Survey of London. Volume 25: St. George's Fields (The Parishes of St. George the Martyr Southwark and St. Mary Newington)* (London: London County Council, 1955), pp. 39–48; 50 Geo. III, c. 191 (1810); see also 52 Geo. III, c. 211(1812).

77 LMA, P97/MRY/050, St. Mary Magdalene, Woolwich vestry minutes, 1768–1819, cited in *Survey of London, Vol. 48*, pp. 419–20; S. Newsome and A. Williams, *Woolwich Common, Woolwich, Greater London: An Assessment of the Historic Environment of Woolwich Common and its Environs* (Swindon:

English Heritage Research Department Report, 98, 2009), pp. 7–8; 42 Geo. III, c. 89; 43 Geo. III, c. 35 and 44 Geo. III, c. 107.

78 John Mortimer, *The Whole Art of Husbandry; or, the Way of Managing and Improving of Land* (1707), p. 184; *The Complete Grazier: or, Gentleman and Farmer's Directory* (2nd edn, 1767), p. 166; William Marshall, *Minutes of Agriculture; with Experiments and Observations Concerning Agriculture and the Weather* (1783), 5 November 1775.

79 OBP: January 1785, William Tebay (t17850112–21); February 1785, Thomas Hamilton and Simon Goring (t17850223–95).

80 C. Shammas, *The Pre-Industrial Consumer in England and America* (Oxford: Clarendon, 1990), pp. 29–30, 302 and Table 2.2.

81 OBP: January 1774, William Archer (t17740112–34); April 1824, John Thomas, Thomas Oakes, Daniel Castle (t18240407–7); May 1823, Thomas Quin (t18230514–80).

82 OBP: September 1752, James Penprice and Edward Perry (t17520914–26); October 1773, Michael Smith (t17731020–2); September 1797, Henry Shippey (t17970920–69); December 1798, William Longford (t17981205–16); September 1818, John Jones and James Crouch (t18180909–277); April 1828, Benjamin Spencer (t18280410–260).

83 Thomas Bell, *A History of British Quadrupeds, Including the Cetacea* (1837), p. 363.

84 OBP, October 1773, Michael Smith (t17731020–2).

85 OBP, February 1790, Elizabeth Taylor, Benjamin Franklin (t17900224–68); June 1796, Edward Montague (t17960622–15); Stow [updated by Strype], *A Survey of London*, vol. 2, p. 343.

86 OBP, February 1785, Thomas Hamilton, Simon Goring (t17850223–95); February 1794, William Sullivan (t17940219–77).

87 OBP: July 1730, Edward Shaftoe (t17300704–45); September 1752, James Penprice and Edward Perry (t17520914–26); October 1785, John Wood (t17851019–52); September 1797, Henry Shippey (t17970920–69); January 1814, Thomas Raven (t18140112–102); February 1829, George Humphries (t18290219–12).

88 OBP: January 1754, William Irons, Benjamin Richford (t17540116–38); April 1788, Thomas Matthews (t17880402–40); February 1790, Elizabeth Taylor, Benjamin Franklin (t17900224–68); February 1794, William Sullivan (t17940219–77); June 1796, Edward Montague (t17960622–15).

89 OBP: June 1796, Edward Montague (t17960622–15); January 1811, John Rayson (t18110109–75).

90 OBP, May 1803, James Rudd, John Willis (t18030525–38).

91 OBP: December 1827, John Crudgington, David Baron (t18271206–196); October 1785, John Wood (t17851019–52); June 1796, Edward Montague (t17960622–15); December 1826, Robert Gee (t18261207–225).

92 Malcolmson, 'Getting a living in the slums of Victorian Kensington', 34.

93 Gavin, *Sanitary Ramblings*, p. 111; Friedrich Engels, *The Condition of the Working Class in England*, ed., V. Kiernan (1844; Harmondsworth: Penguin, 1987), pp. 91–2, 124–5; R. Scola, *Feeding the Victorian City: The Food Supply of Manchester, 1770–1870*, ed. W. A. Armstrong and P. Scola (Manchester: Manchester University Press, 1992), p. 39.

94 B. M. Short, 'The south east: Kent, Surrey and Sussex', in Thirsk (ed.), *The Agrarian History of England and Wales, vol. 5, 1640–1750: 1. Regional Farming Systems*, p. 307.

95 LMA, P92/SAV/0444, A survey and valuation.

96 OBP, February 1825, William Sullivan, Joseph Flowers, William Page (t18250217–179); LL, WACWIC652160463 and Westminster Abbey Muniment Room, Westminster Coroners' Inquest into the death of Richard Pidgeon, 1776.

97 Surrey History Centre, QS2/6/1784/Mic/57, Surrey Quarter Sessions, Sessions Bundles, Eleanor Harris, The Information of Joseph Whiting Holmes of the parish of Saint George Southwark… Cow Keeper, 19 July 1784.

98 OBP, December 1773, Benjamin Martin, John Ridley (t17731208–49); Southwark Local History Library, 2536, Vestry Minutes, Parish of St Giles, Camberwell, 29 October–5 November 1789.

99 LMA, COL/SP/05/084, Order of Court of Aldermen that persons who bring in stray cattle shall give their true name and address the Keeper of the Green-Yard and shall not be rewarded until 48 hours later; to prevent vagrant persons from driving cattle from fields to the Green Yard and giving fictitious names, 14 March 1731/2.

100 Thomas Baird, *General View of the Agriculture of the County of Middlesex: With Observations on the Means of its Improvement* (1793), p. 13; Robert Pollard after Edward Dayes, *View of Bloomsbury Square* (etching and aquatint, 1787); [Legg], *Low-Life*, p. 27; BMDPD, Heal, 4.6, Anon., Draft trade card of Martha Prockter and Lydia Edwards, ass-keepers (verso: a bill dated 1775).

101 *Sunday Reformer and Universal Register* (30 June 1793); C. Plumb, 'Exotic animals in eighteenth-century Britain' (PhD thesis, University of Manchester, 2010), pp. 74–5.

102 George Scharf, *The Westminster Dairy in the Quadrant, Regent Street, near Piccadilly Circus* (drawing, 1825), British Museum.

103 Valenze, *Milk*, p. 116.
104 Thomas Bowles III, *A New and Exact Prospect of the North Side of the City of London taken from the Upper Pond near Islington* (Etching and engraving, 1730).
105 Charles White, after anon., *View of the River Thames from the End of Chelsea* (engraving, 1794).
106 Charles Jenner, *Town Eclogues* (1772), p. 27.
107 Horace Walpole to George Montagu, 6 June 1752, in Horace Walpole, *The Letters of Horace Walpole, Earl of Orford: Including Numerous Letters now First Published from the Original Manuscripts*, ed. anon., 6 vols (1840), p. 425.
108 Anon., *The Red Cow's Speech, to a Milk-Woman, in St. J----S's P----K* (early 1700s).
109 Grosley, *A Tour to London*, vol. 1, p. 87.
110 Sylas Neville, diary entry for 3 August 1767, in B. Cozens-Hardy (ed.), *The Diary of Sylas Neville, 1767–1788* (London: Oxford University Press, 1950), p. 22.
111 William Hogarth, *The Four Times of the Day: Evening* (oil on canvas, 1736), private collection; M. Hallett and C. Riding (eds), *Hogarth* (London: Tate Publishing, 2006), p. 135.

4 Meat on the hoof

1 Jenner, *Town Eclogues*, pp. 26–7.
2 R. S. Metcalfe, *Meat, Commerce and the City: The London Food Market, 1800–1855* (London: Pickering & Chatto, 2012); I. Maclachlan, 'A bloody offal nuisance: The persistence of private slaughter-houses in nineteenth-century London', *Urban History*, 34 (2007), 227–54; Donald, '"Beastly sights"', 530–40; C. Philo, 'Animals, geography, and the city: notes on inclusions and exclusions', in Wolch and Emel (eds), *Animal Geographies*, pp. 61–1; P. Joyce, 'Maps, blood and the city: The governance of the social in nineteenth-century Britain', in P. Joyce (ed.), *The Social in Question: New Bearings in History and the Social Sciences* (London: Routledge, 2002), pp. 97–114.
3 Dodd, *The Food of London*, p. 233.
4 N. McKendrick, J. Brewer and J. H. Plumb, *The Birth of a Consumer Society: The Commercialization of Eighteenth-Century England* (London: Europa, 1982); B. Lemire, *Fashion's Favourite: The Cotton Trade and the Consumer in Britain, 1660–1800* (Oxford: Oxford University Press, 1991); J. Brewer and R. Porter (eds), *Consumption and the World of Goods* (London: Routledge, 1993);

S. Richards, *Eighteenth-Century Ceramics: Products for a Civilised Society* (Manchester: Manchester University Press, 1999); W. D. Smith, *Consumption and the Making of Respectability 1600–1800* (London: Routledge, 2002); M. Berg, *Luxury and Pleasure in Eighteenth-Century Britain* (Oxford: Oxford University Press, 2005); J. Styles and A. Vickery (eds), *Gender, Taste, and Material Culture in Britain and North America, 1700–1830* (New Haven and London: Yale University Press, 2006); F. Trentmann (ed.), *The Oxford Handbook of the History of Consumption* (Oxford: Oxford University Press, 2012).

5 J. A. Chartres, 'City and towns, farmers and economic change in the eighteenth century', *Historical Research*, 64 (1991), 138–55 and 'Food consumption and internal trade', in A. L. Beier and R. Finlay (eds), *London, 1500–1700: The Making of the Metropolis* (London: Longman, 1986), pp. 168–96; Reed, 'London and its hinterland', pp. 65–6; W. Thwaites, 'Oxford food riots: A community and its markets', in A. Randall and A. Charlesworth (eds), *Markets, Market Culture and Popular Protest in Eighteenth-Century Britain and Ireland* (Liverpool: Liverpool University Press, 1996), pp. 137–62; Mingay (ed.), *The Agrarian History of England and Wales*, vol. 6.

6 Cronon, *Nature's Metropolis*, pp. 255–6.

7 A. Saunders (ed.), *The Royal Exchange* (London: London Topographical Society, 1997); D. Hancock, *Citizens of the World: London Merchants and the Integration of the British Atlantic Community, 1735–1785* (Cambridge: Cambridge University Press, 1995); Mokyr, *The Enlightened Economy*, ch. 11; H. C. Mui and L. H. Mui, *Shops and Shopkeeping in Eighteenth-Century England* (London: Routledge, 1989); C. Walsh, 'Shop design and the display of goods in eighteenth-century London', *Journal of Design History*, 8 (1995), 157–76. On the importance of markets, see C. Smith, 'The wholesale and retail markets of London, 1660–1840', *EcHR*, 55 (2002), 31–50; Scola, *Feeding the Victorian City*.

8 John Houghton, *Husbandry and Trade Improv'd: Being a Collection of Many Valuable Materials Relating to Corn, Cattle, Coals, Hops, Wool &c.*, revised by Richard Bradley, 3 vols (1727), pp. 300–1, 314.

9 Nightingale, *London and Middlesex*, p. 476; Dodd, *The Food of London*, pp. 249–50.

10 'The Clerk of the Market's Account for the Year 1725', in Maitland, *The History and Survey of London* (1760), vol. 2, p. 756; John Christian Curwen, *Hints on Agricultural Subjects and on the Best Means of Improving the Condition of the Labouring Classes* (2nd edn, 1809), p. 131.

11 Young (ed.), *Annals of Agriculture* (1784), vol. 2, pp. 420–1.

12 R. Perren, 'Markets and marketing', in Mingay (ed.), *The Agrarian History of England and Wales*, vol. 6, pp. 192–3.

13 LMA, CLA/016/AD/02/006, Second report of the Select Committee of the House of Commons appointed to inquire into the state of Smithfield Market, and the slaughtering of cattle in the metropolis (1828), pp. 95–7; Rodger, *The Wooden World*, pp. 70–86.

14 Middleton, *View of the Agriculture of Middlesex*, p. 409; Bell, *A System of Geography*, vol. 3, p. 102.

15 Middleton, *View of the Agriculture of Middlesex*, pp. 409–11; Guildhall Library, Closed Access Broadside 30.74, S. T. Jannsen, *A Table Shewing the Number of Sheep and Black Cattle Brought to Smithfield Market for the Last Forty Years* (undated).

16 Mingay (ed.), *The Agrarian History of England and Wales*, vol. 6, p. 943; J. Broad, 'Cattle plague in eighteenth-century England', *AgHR*, 31 (1983), 114; L. Wilkinson, *Animals and Disease: An Introduction to the History of Comparative Medicine* (Cambridge: Cambridge University Press, 1992), pp. 35–65.

17 *The Parliamentary Debates from the Year 1803 to the Present Times*, vol. 26 (1812), p. 399.

18 R. Horowitz, J. M. Pilcher and S. Watts, 'Meat for the multitudes: Market culture in Paris, New York City, and Mexico City over the long nineteenth century', *The American Historical Review*, 109 (2004), 1060–1.

19 David Hughson, *London; Being an Accurate History and Description of the British Metropolis and its Neighbourhood: To Thirty Miles Extent, from an Actual Perambulation*, 6 vols (1809), vol. 6, p. 601; Bell, *A System of Geography*, vol. 3, p. 102.

20 Maclachlan, 'A bloody offal nuisance', 228, 234; Dodd, *The Food of London*, pp. 249–50.

21 Cronon, *Nature's Metropolis*, pp. 207–11.

22 Adam Anderson, *An Historical and Chronological Deduction of the Origin of Commerce, From the Earliest Accounts*, 4 vols (1789), vol. 4, pp. 660–1.

23 *The Universal Magazine*, vol. 11 (1809), p. 75; Hughson, *London*, vol. 4, p. 597; LMA, CLA/016/AD/01/003, Substance of the Bill Now Before Parliament, For Enlarging and Improving the Market Place with Observations Thereon: Also the Objections to the Proposed Measure, and Answer Thereto (1813); Nightingale, *London and Middlesex*, p. 476; Bell, *A System of Geography*, vol. 3, p. 102.

24 Mokyr, *The Enlightened Economy*, pp. 222–4; L. Brunt, 'Rediscovering risk: Country banks as venture capital firms in the first Industrial Revolution', *Journal of Economic History*, 66 (2006), 74–102.

25 Metcalfe, *Meat, Commerce and the City*, p. 30; Dodd, *The Food of London*, p. 237; LMA, CLA/016/AD/02/006, Second report, p. 206; LMA, CLA/016/AD/01/003, Substance of the Bill.

26 LMA, CLA/016/AD/02/006, Second report, Introduction and pp. 71, 146; LMA, CLA/016/FN/01/007, Rough weekly account: Tolls collected at Smithfield Market and the City Gates, 2 September 1727–28 September 1728; LMA, CLA/016/FN/01/004, Dues Collected at Smithfield Market, 28 March 1777–31 December 1817.

27 LMA, CLA/016/AD/02/006, Second report, appendix and p. 62; Bell, *A System of Geography*, vol. 3, p. 102; Middleton, *View of the Agriculture of Middlesex*, pp. 409–11.

28 4 & 5 Henry VII, c. 3 (1488), repealed by 24 Henry VIII, c. 16 (1532).

29 Lawrence, *A Philosophical and Practical Treatise on Horses*, vol. 1, p. 156.

30 Scola, *Feeding the Victorian City*, pp. 46, 150, 156; Richard Brooke, *Liverpool as it was During the Last Quarter of the Eighteenth Century, 1775 to 1800* (Liverpool, 1853), p. 117; *Preston Chronicle* (11 June 1831 and 2 July 1831); LMA, CLA/016/AD/02/006, Second report, appendix, evidence provided by the Chamberlain's Office.

31 Horowitz, Pilcher and Watts, 'Meat for the multitudes', 1061; S. Watts, 'Boucherie et hygiène à Paris au XVIIIe siècle', *Revue d'histoire moderne et contemporaine*, 51 (2004), 79–103; S. Watts, *Meat Matters: Butchers, Politics, and Market Culture in Eighteenth-Century Paris* (Rochester, NY: University of Rochester Press, 2006).

32 LMA, CLA/016/AD/02/006, Second report, p. 85, evidence of Charles Whitlaw, agriculturalist and botanist; E. G. Burrows and M. Wallace, *Gotham: A History of New York City to 1898* (Oxford: Oxford University Press, 1999), pp. 347, 355, 475, 658.

33 P. E. Jones, *The Butchers of London: A History of the Worshipful Company of Butchers of the City of London* (London: Secker & Warburg, 1976), p. 103; Hughson, *London*, vol. 6, p. 598.

34 29 Geo. II, c. 88 (1756); *Whitehall Evening Post* (31 July 1755); *Gazetteer and London Daily Advertiser* (13 August 1756); *Whitehall Evening Post* (18 September 1756); F. H. W. Sheppard, *Local Government in St Marylebone, 1688–1835: A Study of the Vestry and the Turnpike Trust* (London: University of London, Athlone Press, 1958), pp. 94–101.

35 Parliamentary Archives, HL/PO/JO/10/3/250/5, Islington to Paddington Road Bill, Petition of Saint George Hanover Square, St James Westminster, Saint Ann Soho, Paddington and Saint Marylebone, 6 April 1756.

36 Parliamentary Archives, HL/PO/JO/10/3/250/4, Islington to Paddington Road Bill, Petition of Saint Andrews Holborn, Saint Georges Bloomsbury and Saint Giles in the Fields, 6 April 1756.

37 Jonas Hanway, *A Letter to Mr. John Spranger on his Excellent Proposal for Paving, Cleansing, and Lighting the Streets of Westminster, and the Parishes in*

Middlesex (1754); [Joseph Massie], *An Essay on the Many Advantages Accruing to the Community from the Superior Neatness, Conveniences, Decorations and Embellishments of Great and Capital Cities* (1754), pp. 12–15.

38 Parliamentary Archives, HL/PO/JO/10/3/250/14, Islington to Paddington Road Bill, Petition of the Graziers, Salesmen, Butchers, Drovers and Dealers in Cattle who Attend Smithfield Market, 9 April 1756.

39 Hughson, *London*, vol. 6, p. 600.

40 *The Gentleman's Magazine*, 28 (1758), p. 92.

41 S. Roberts, *The Story of Islington* (London: The Crowood Press, 1975), pp. 55, 175; Atkins, 'The milk trade of London', p. 28; LMA, CLA/016/AD/02/006, Second report, p. 235.

42 *Journal of the House of Commons, vol. 61: 1806* (1806), p. 230; LMA, CLA/016/AD/02/006, Second report, p. 155.

43 Nightingale, *London and Middlesex*, p. 479; LMA, CLA/016/AD/02/006, Second report, pp. 46, 138; Maclachlan, 'A bloody offal nuisance', 227.

44 Hughson, *London*, vol. 6, p. 599.

45 H-J. Voth, 'Time and work in eighteenth-century London', *Journal of Economic History*, 58 (1998), 32–6.

46 *Lloyd's Evening Post* (19 August 1765); *St James's Chronicle* (20 August 1793); *Morning Chronicle* (25 July 1808 and 23 December 1818).

47 LMA, CLA/016/AD/02/006, Second report, p. 144.

48 Dodd, *The Food of London*, p. 234.

49 LMA, CLA/016/AD/02/011, Substance of the Cutting Butcher's Petition and Allegations Offered for an Alteration of Smithfield Market from Friday to Thursday (1796).

50 CLA/016/02/006, Second report, pp. 171–6; Jones, *The Butchers of London*, p. 101.

51 I. Ekesbo, *Farm Animal Behaviour: Characteristics for Assessment of Health and Welfare* (Wallingford: CABI, 2011), pp. 82–5; A. F. Fraser, *Farm Animal Behaviour: An Introductory Textbook on the Study of Behaviour as Applied to Horses, Cattle, Sheep and Pigs* (London: Baillière Tindall, 1974), p. 64; J. J. Lynch, G. N. Hinch and D. B. Adams, *The Behaviour of Sheep: Biological Principles and Implications for Production* (Wallingford: CABI, 1992), p. 94.

52 LMA, CLA/016/AD/02/006, Second report, pp. 16, 25, 164.

53 LMA, CLA/016/AD/02/006, Second report, pp. 44, 220.

54 H. Velten, *Cow* (London: Reaktion Books, 2007), p. 20.

55 John Gwynn, *London and Westminster Improved, Illustrated by Plans: To which is Prefixed, a Discourse on Publick Magnificence* (1766), p. 19.

56 OBP, October 1786, Thomas Plata, Francis Parker (t17861025–37).

57 Youatt, *Cattle*, pp. 164, 274, 283.

58 OBP, October 1786, Thomas Plata, Francis Parker (t17861025–37).

59 *London Evening Post* (7 May 1757); *Public Advertiser* (6 March 1767); *London Chronicle* (26 December 1789).

60 PAL, HL/PO/JO/10/3/250/14, Islington to Paddington Road Bill, Petition of the Graziers; see also Hughson, *London*, vol. 6, p. 600.

61 OBP, October 1786, Thomas Plata, Francis Parker (t17861025–37); LMA, CLA/016/AD/02/006, Second report, p. 147.

62 *The Annual Register, Or a View of the History, Politics and Literature, For the Year 1761* (6th edn, 1796), p. 106. See also *Evening Post* (13 May 1763); *Lloyd's Evening Post* (27 May 1764); *St James's Chronicle* (12 May 1769).

63 LMA, CLA/016/AD/02/006, Second report, p. 155.

64 Gray, *Crime, Prosecution and Social Relations*, p. 118.

65 OBP, October 1786, Thomas Plata, Francis Parker (t17861025–37).

66 Anon., *A Dissertation on Mr. Hogarth's Six Prints Lately Published, viz. Gin-Lane, Beer-Street, and the Four Stages of Cruelty* (1751), p. 37.

67 M. Thale (ed.), *The Autobiography of Francis Place, 1771–1854* (Cambridge: Cambridge University Press, 1972), pp. 69–70.

68 21 Geo. III, c. 67 (1781); CLA/015/AD/02/032, Warrants for payments to constables and others for the apprehension and prosecution of persons, not being employed to drive cattle, for the 'hunting away' of bullocks, October–December 1789.

69 D. T. Andrew, *Philanthropy and Police: London Charity in the Eighteenth Century* (Princeton: Princeton University Press, 1989), p. 163.

70 LMA, MJ/O/C/006, Middlesex Sessions, General orders of court, 6 September 1753 (LL, LMSMGO556030027); *London Chronicle* (19 October 1762); *Craftsman* (18 July 1772).

71 Gray, *Crime, Prosecution and Social Relations*, p. 141.

72 *Gazetteer* (10 November 1764).

73 M. Horkheimer and T. W. Adorno, *Dialectic of Enlightenment: Philosophical Fragments*, ed. G. S. Noerr, trans. E. Jephcott (1947; Stanford, CA: Stanford University Press); R. Porter, 'Accidents in the eighteenth century', in R. Cooter and B. Luckin (eds), *Accidents in History: Injuries, Fatalities and Social Relations* (Amsterdam: Rodopi, 1997), p. 97; S. Vogel, *Against Nature: The Concept of Nature in Critical Theory* (Albany: State University of New York Press, 1996), pp. 51–68; Drayton, *Nature's Government*.

74 DeJohn Anderson, *Creatures of Empire*, pp. 96, 132.

75 *St James's Chronicle* (17 October 1761); *Public Advertiser* (17 April 1765); *Common Sense* (14 April 1739); *London Daily Advertiser* (6 February 1752).

76 Beattie, *Crime and the Courts*, p. 89.

77 *Gazetteer* (17 November 1764); *London Chronicle* (23 March 1765).
78 Thomas Rowlandson, *The Overdrove Ox* (etching and aquatint, 1787).
79 21 Geo. III, c. 67 (1781); K. J. Bonser, *The Drovers: Who They Were and How They Went: An Epic of the English Countryside* (London: Macmillan, 1970), p. 88.
80 *Public Advertiser* (9 November 1792).
81 *London Packet or New Lloyd's Evening Post* (10–13 May 1793); *Morning Herald* (16 January 1793); OBP, July 1824, Thomas Abrahams (t18240715–128); *The Voice of Humanity* (1827), p. 26 and (1830), p. 108.
82 Burney Collection.
83 *Whitehall Evening Post* (10 January 1758 and 1 November 1759).
84 Donald, '"Beastly sights"', 530–40; J. White, *London in the Nineteenth Century: A Human Awful Wonder of God* (London: Vintage, 2008), pp. 188–9; Kean, *Animal Rights*, pp. 58–64; Philo, 'Animals, geography, and the city', 60–7; Maclachlan, 'A bloody offal nuisance', 231; Metcalfe, *Meat, Commerce and the City*.
85 James Mathews, *Remarks on the Cause and Progress of the Scarcity and Dearness of Cattle Swine, Cheese, &c. and of the articles tallow candles and soap &c.* (1797), pp. 87–8.
86 Maclachlan, 'A bloody offal nuisance'.
87 Ogborn, *Spaces of Modernity*, pp. 21, 236.
88 Henry Fielding, *The Grub-Street Opera* (1731), p. 41; 'The Roast Beef of Old England. A Cantata. Taken from the Celebrated Print of the Ingenious Mr Hogarth', in *The Bull-Finch. Being a Choice Collection of the Newest and Most Favourite English Songs Which Have Been Sett to Music and Sung at the Public Theatres & Gardens* (1760), pp. 99–103; William Hogarth, *O the Roast Beef of Old England ('The Gate of Calais')* (oil on canvas, 1748), Tate; Boswell, *The Life of Samuel Johnson*, pp. 217, 663–4; E. V. Roberts, 'Henry Fielding and Richard Leveridge: Authorship of "The Roast Beef of Old England"', *Huntingdon Library Quarterly*, 27 (1964), 175–81; O. Baldwin and T. Wilson, '250 years of roast beef', *The Musical Times* (April 1985), 203–7; B. Rogers, *Beef and Liberty: Roast Beef, John Bull and the English Nation* (London: Chatto and Windus, 2003); Ritvo, *The Animal Estate*, p. 390.
89 OBP: April 1778, Charles Atkins (t17780429–70); January 1790, Ann Harney, Ann Jackson (t17900113–2); January 1807, John Fordham, John Harvey, Richard Hartford, William Bridge (t18070114–5); July 1812, Charles Harding (t18120701–48).
90 Culley, *Observations on Live Stock*, p. 226.

91 Zachariah Allen, *Sketches of the State of the Useful Arts, and of Society, Scenery, &c. &c. in Great Britain, France and Holland. Or, the Practical Tourist*, 2 vols (Boston, 1833), vol. 2, p. 297.

92 Culley, *Observations on Live Stock*, p. 61.

93 Holderness, 'Prices, productivity, and output', p. 154; Edward Lisle, *Observations in Husbandry*, 2 vols (2nd edn, 1757), vol. 2, p. 14.

94 Daniel Defoe, *A Tour Thro' the Whole Island of Great Britain*, 4 vols (3rd edn, 1742), vol. 1, p. 304; *The Beauties of All the Magazines Selected, For the Year 1764*, vol. 3 (1764), p. 358; Bath and West of England Society, *Letters and Papers on Agriculture, Planting &c., Selected from the Correspondence of the Bath and West of England Society, for the Encouragement of Agriculture, Arts, Manufactures and Commerce*, vol. 7 (Bath, 1795), p. 226.

95 Cobbett, *Rural Rides*, p. 372.

96 *Morning Chronicle* (20 May 1775).

97 *Whitehall Evening Post* (11–13 February 1794).

98 *London Chronicle or Universal Evening Post* (9–12 March 1765); *London Evening Post* (23–25 April 1765).

99 *General Remark on Trade* (11–13 August 1707); *Whitehall Evening Post* (21–24 June 1788); *Sun* (28 January 1794); Maitland, *The History and Survey of London* (1760), vol. 2, pp. 755–7; Feltham, *The Picture of London* (1813), p. 73; Hughson, *London*, vol. 6, p. 399; Nightingale, *London and Middlesex*, p. 469; Bell, *A System of Geography*, vol. 3, p. 102; Youatt, *Cattle*, pp. 256–7.

100 Middleton, *View of the Agriculture of Middlesex*, pp. 409–12; Guildhall Library, Closed Access Broadside 30.74, A Table Shewing the Number of Sheep and Black Cattle.

101 A. Savile (ed.), *Secret Comment, The Diaries of Gertrude Savile, 1721–1759* (Nottingham: Thoroton Society, 1997), pp. 258, 270. See also Thomas, *Man and the Natural World*, p. 182.

102 *Westminster Journal or New Weekly Miscellany* (14 February 1747).

103 Wilkinson, *Animals and Disease*, pp. 35–65; Broad, 'Cattle plague', 105.

104 Savile (ed.), *The Diaries of Gertrude Savile*, p. 290.

105 Walter Arnold, *The Life and Death of the Sublime Society of Beef Steaks* (1871); R. J. Allen, *The Clubs of Augustan London* (Cambridge, MA: Harvard University Press, 1933), pp. 137–45.

106 *World* (21 October 1790); R. Blake, *George Stubbs and the Wide Creation: Animals, People and Places in the Life of George Stubbs, 1724–1806* (London: Pimlico, 2005), pp. 252–3; Ritvo, *The Animal Estate*, pp. 45–6; Rogers, *Beef and Liberty*, p. 175.

107 E. J. Powell, *History of the Smithfield Club, 1798 to 1900* (London: The Smithfield Club, 1902).

108 Maclachlan, 'A bloody offal nuisance', p. 237.
109 Thomas, *Man and the Natural World*, p. 182.

5 Consuming horses

1 L. Weatherill, *Consumer Behaviour and Material Culture in Britain 1660–1760* (London: Routledge, 1988), ch. 4; P. Borsay, 'The London connection: Cultural diffusion and the eighteenth-century provincial town', *London Journal*, 19 (1994), 21–35; Barnett, *London, Hub of the Industrial Revolution*, ch. 4; H. Berry, 'Prudent luxury: The metropolitan tastes of Judith Baker, Durham gentlewoman', in R. Sweet and P. Lane (eds), *Women and Urban Life in Eighteenth-Century England* (Aldershot: Ashgate, 2003), pp. 131–56; Greig, *The Beau Monde*, ch. 1.
2 McKendrick, Brewer and Plumb, *The Birth of a Consumer Society*; Weatherill, *Consumer Behaviour and Material Culture*; Lemire, *Fashion's Favourite*; Brewer and Porter (eds), *Consumption and the World of Goods*; Richards, *Eighteenth-Century Ceramics*; Smith, *Consumption and the Making of Respectability*; Berg, *Luxury and Pleasure*; Styles and Vickery (eds), *Gender, Taste, and Material Culture*.
3 E. Kowaleski-Wallace, *Consuming Subjects: Women, Shopping, and Business in the Eighteenth Century* (New York: Columbia University Press, 1997); C. Walsh, 'The design of London goldsmiths' shops in the early eighteenth century' and 'Shop design and the display of goods', in D. Mitchell (ed.) *Goldsmiths, Silversmiths and Bankers: Innovation and the Transfer of Skill, 1550–1750* (London: Centre for Metropolitan History, 1995) pp. 96–111, 157–76; J. Stobart, 'Shopping streets as social space: Leisure, consumerism and improvement in an eighteenth-century county town', *Urban History*, 25 (1998), 3–21; A. Hann and J. Stobart, 'Sites of consumption: The display of goods in provincial shops in eighteenth-century England', *Journal of the Social History Society*, 2 (2005), 165–87; H. Berry, 'Polite consumption: Shopping in eighteenth-century England', *Transactions of the Royal Historical Society*, 12 (2002), 375–94.
4 Lord Pembroke to Lord Herbert, Wilton House, 6 April 1781, in S. Herbert (ed.), *Pembroke papers, 1780–1794: Letters and Diaries of Henry, Tenth Earl of Pembroke and his Circle* (London: Jonathan Cape, 1950), pp. 111–12.
5 John Lawrence, *A Philosophical and Practical Treatise on Horses*, vol. 2, pp. 168, and *The Horse in All His Varieties and Uses; His Breeding, Rearing, and Management, Whether in Labour or Rest; with Rules, Occasionally Interspersed, for His Preservation from Disease* (1829), pp. 147–9.
6 *World* (29 March 1788 and 22 October 1788).

7 Youatt, *The Horse*, p. 29. See also D. Landry, *Noble Brutes: How Eastern Horses Transformed English Culture* (Baltimore: Johns Hopkins University Press, 2008), pp. 1–6.

8 John Sinclair, *The History of the Public Revenue of the British Empire, Part III* (1790), p. 251.

9 Feltham, *The Picture of London* (1825), p. 352; Nightingale, *London and Middlesex*, p. 479; LMA, CLA/016/AD/02/006, Second report, pp. 218–19.

10 Samuel Pegge (ed.), *Fitz-Stephen's Description of the City of London, Newly Translated from the Latin Original; with a Necessary Commentary… By an Antiquary* (1772), pp. 36–7

11 G. Robinson, *Horses, People and Parliament in the English Civil War: Extracting Resources and Constructing Allegiance* (Farnham: Ashgate Publishing, 2012), pp. 193–201; P. Edwards, 'The supply of horses to the parliamentarian and royalist armies in the English Civil War', *Historical Research*, 68 (1995), 49–66; Defoe, *A Tour* (1986), p. 313.

12 LMA, CLA/016/AD/02/006, Second report, p. 218.

13 William Gaspey, *Tallis's Illustrated London; In Commemoration of the Great Exhibition of all Nations in 1851*, 2 vols (1851), vol. 1, p. 231; Charles Knight, *Knight's Cyclopaedia of London* (1851), p. 798.

14 LMA, CLA/016/AD/02/006, Second report, p. 219.

15 Lawrence, *The Horse in All His Varieties*, p. 155. See also Theodore Lane, 'How to Pick up a "RUM ONE to Look at" and a "GOOD ONE to Go" in Smithfield' (hand-coloured etching and aquatint, 1825).

16 Adam-Wolfgang Töppffer quoted in R. Loche, C. Sanger, L. Boissonnas, *Jacques-Laurent Agasse 1767–1849*, exhibition catalogue (London: Tate Gallery, 1988), p. 264.

17 J. A. Chartres, 'The capital's provincial eyes: London's inns in the early eighteenth century', *London Journal*, 3 (1977), 27 and 'The eighteenth-century English inn', p. 220.

18 *Public Advertiser* (6 July 1754).

19 Burney Collection; *Daily Courant* (19 July 1703).

20 Burney Collection; S. Morison, *The English Newspaper: Some Account of the Physical Development of Journals Printed in London Between 1622 and the Present Day* (Cambridge: Cambridge University Press, 1932; reprinted 2009), ch. 9; J. Black, *The English Press in the Eighteenth Century* (London: Croom Helm, 1987), p. 42.

21 Feltham, *The Picture of London* (1818), pp. 417–18.

22 Lawrence, *The Horse in All His Varieties*, p. 152; *Daily Advertiser* (3 March 1778); *Morning Post & Daily Advertiser* (7 March 1785); *Morning Post and Fashionable World* (28 August 1795);

23 LMA, MS 11936/373, Sun policy 584787, Joseph Aldridge (1792) and MS 11936/389, Sun policy 609420, Edmund Tattersall (1792).

24 V. Orchard, *Tattersalls: Two Hundred Years of Sporting History* (London: Hutchinson, 1953), pp. 140, 168. See also Robert Dighton, *Two Impures of the Ton driving to the Gigg Shop, Hammersmith* (mezzotint, 1781); Thomas Holcroft, *The Road to Ruin: A Comedy. As it is Acted at the Theatre-Royal, Covent-Garden* (1792), p. 31; J. G. Holman, *Abroad and at Home. A comic Opera, in Three Acts. Now Performing at the Theatre-Royal, Covent-Garden* (1796), p. 28; George Parker, *A View of Society and Manners in High and Low Life* (1781), p. 48; James Elmes, *Metropolitan Improvements: Or, London in the Nineteenth Century, Displayed in a Series of Engravings of the New Buildings, Improvements, &c. by the Most Eminent Artists, from Original Drawings, taken from the Objects themselves expressly for this Work* (1827–30), pp. 142–3.

25 Mui and Mui, *Shops and Shopkeeping*, p. 239; J. G. Carrier, *Gifts and Commodities: Exchange and Western Capitalism Since 1700* (London: Routledge, 1995), p. 81.

26 Anon., *Interior View of the Auction Rooms at Aldridge's Horse Repository* (lithograph, 1824); Elmes, *Metropolitan Improvements*, pp. 142–3.

27 Lawrence, *The Horse in All His Varieties*, pp. 149–52; R. Longrigg, *The English Squire and his Sport* (London: Michael Joseph, 1977), p. 201; M. Huggins, *Flat Racing and British Society 1790–1914: A Social and Economic History* (London: Frank Cass, 2000), pp. 20–22, 57–9; W. Vamplew and J. Kay (eds), *Encyclopaedia of British Horse Racing* (London: Routledge, 2005), p. 304.

28 OBP: February 1773, Francis Mercier (t17730217–2); April 1774, Charles Green (t17740413–11); September 1779, Ann Lascelles (t17790915–1); John Trusler, *The London Advisor and Guide: Containing Every Instruction and Information Useful and Necessary to Persons Living in London, and Coming to Reside There* (1786), p. 171.

29 Lambeth Archives, P/S/13/19, Assessments for taxes on houses and windows; inhabited houses, male servants; four-wheeled carriages; riding and carriage horses; horses used in husbandry and trade; mules, 1800–01.

30 Trusler, *The London Advisor and Guide*, pp. 169–73.

31 OBP: September 1763, George Wilkins (t17630914–19); April 1765, William Hornsby (t17650417–37); April 1765, Joseph Steel (t17650417–15); December

1774, John Rooke (t17741207–61). On these items, see also A. Vickery, 'His and hers: Gender, consumption and household accounting in eighteenth-century England', *Past & Present*, supplement 1 (2006), 27–8, 33–4.

32 Barnett, *London, Hub of the Industrial Revolution*, p. 85; Pigot & Co, *Metropolitan New Alphabetical Directory for 1827* (1827).

33 LL.

34 OBP, June 1767, Robert Johnson (t17670603–41); Trusler, *The London Advisor and Guide*, p. 170.

35 M. H. MacKay, 'The rise of a medical specialty: The medicalisation of equine care *c.* 1680–*c.* 1800' (PhD thesis, University of York, 2009), pp. 170–2, 234.

36 Sheffield Archives, Wentworth Woodhouse Muniments, WWM/A/1300, Weekly abstract, tradesmen's bills on house and stables, London, 1775–82.

37 West Sussex County Records Office, Petworth House Archives, PHA 7555, Abstract, observations, and receipted bills from London tradesmen to 3rd Earl of Egremont, 1797–98; PHA 6640, Abstract of London tradesmen's bills paid for 3rd Earl of Egremont, 1794–95; PHA 6635, London tradesmen's bills from 25 June 1791 to 25 June 1792; MacKay, 'The rise of a medical specialty', pp. 91–100, 152–8, 175, 186–204.

38 Porter, *London*, p. 145; R. Stewart, *The Town House in Georgian London* (New Haven and London: Yale University Press, Paul Mellon Centre, 2009); Cruickshank and Burton, *Life in the Georgian City*; D. J. Olsen, *Town Planning in London: In the Eighteenth and Nineteenth Centuries* (New Haven: Yale University Press, 1964). A rare exception is G. Worsley, *The British Stable* (New Haven and London: Yale University Press, 2004), ch. 5.

39 Rocque, *A Plan of the Cities of London and Westminster, and Borough of Southwark*; Horwood, *Plan of the Cities of London and Westminster, with the Borough of Southwark*.

40 F. H. W. Sheppard (ed.) *Survey of London. Volume 39: The Grosvenor Estate in Mayfair, Part 1: General History* (London: London County Council, 1977), pp. 11–2; R. McManus, 'Windows on a hidden world: Urban and social evolution as seen from the mews', *Irish Geography*, 37 (2004), 37–40; Worsley, *The British Stable*, p. 110.

41 Sheffield Archives: WWM/A/1228, Inventory and appraisement, 1782–84, pp. 24–7, 'Particulars of horses and carriages &c late the property of the Marquess of Rockingham deceased as valued by Mr. Tattershall', 1–2 August 1782; WWM/A/1228, Inventory & appraisement, 26 August 1782, p. 52.

42 Robert Adam and James Adam, *The Works in Architecture of Robert and James Adam*, 3 vols (1778–1822), vol. 2, plate 1, 'Plan of the Parlour Story, and principal Floor'; Stewart, *The Town House in Georgian London*, p. 97.

43 Maitland, *The History and Survey of London* (1739), pp. 354ff., cited in Earle, *The Making of the English Middle Class*, p. 388, fn. 62.
44 Worsley, *The British Stable*, p. 119.
45 Peter Potter, *Plan of the Parish of St Marylebone in the County of Middlesex* (2nd edn, 1821).
46 H. Greig, 'Leading the fashion: The material culture of London's Beau Monde', in Styles and Vickery (eds), *Gender, Taste, and Material Culture*, p. 299.
47 *World* (21 May 1788).
48 Burney Collection; Potter, *Plan of the Parish of St Marylebone*.
49 J. J. Hecht, *The Domestic Servant Class in Eighteenth-Century England* (London: Routledge, 1956); B. Hill, *Servants: English Domestics in the Eighteenth Century* (Oxford: Clarendon Press, 1996); C. Steedman, *Labours Lost: Domestic Service and the Making of Modern England* (Cambridge: Cambridge University Press, 2009); T. Meldrum, *Domestic Service and Gender, 1660–1750: Life and Work in the London Household* (Harlow: Longman, 2000).
50 Trusler, *The London Advisor and Guide*, p. 184; Sheffield Archives, WWM/A/1296, Isaac Charlton's London household disbursements, 19 July 1781.
51 McShane and Tarr, *The Horse and the City*, p. 127.
52 Mayhew, *London Labour*, vol. 2, pp. 196.
53 Steedman, *Labours Lost*; Adam Smith, *The Wealth of Nations*, ed. A. Skinner (originally published 1776; Harmondsworth: Penguin Classics, 1986), pp. 133–40, 430; E. P. Thompson, *The Making of the English Working Class* (London: Victor Gollancz, 1963).
54 Steedman, *Labours Lost*, pp. 14, 353–4.
55 Voth, 'Time and work in eighteenth-century London', 31–6.
56 OBP, September 1818, George Bays, Alexander Stewart, Mary Stokes (t18180909-23); Johann Wilhelm von Archenholz, *A Picture of England: Containing a Description of the Laws, Customs, and Manners of England. Translated from the French* (1789), p. 125; Sophie von La Roche, *Sophie in London, 1786: Being the Diary of Sophie v. La Roche*, trans. C. Williams (1786; London: Jonathan Cape, 1933), p. 89.
57 William Taplin, *The Sporting Dictionary, and Rural Repository of General Information Upon Every Subject Appertaining to the Sports of the Field*, 2 vols (1803), vol. 1, pp. 349–50.
58 Meldrum, *Domestic Service and Gender*, pp. 124, 167–82; Steedman, *Labours Lost*, pp. 8–9; S. Maza, *Servants and Masters in Eighteenth-Century France: The Uses of Loyalty* (Princeton: Princeton University Press, 1983), pp. 109–10, 131, 134.

59 Sheppard (ed.), *Survey of London. Volume 39*, p. 88.
60 Mayhew, *London Labour*, vol. 1, p. 362.
61 Jonathan Swift, *Directions to Servants in General* (1745; London: Hesperus Press, 2003), pp. 46, 51.
62 Royal Archives, Windsor, Mews/Proc/Mixed, King's Mews Precedence book, 1760–1805, p. 89, 'Orders relative to abuses that have been practiced within the mews', 13 June 1769.
63 Royal Archives, King's Mews Precedence book, pp. 235–6, 'Mr. Smith's Letter to David Parker Esq., King's Mews', 30 April 1789.
64 Steedman, *Labours Lost*, pp. 8–9; P. Earle, *A City Full of People*, p. 85.
65 Meldrum, *Domestic Service and Gender*, p. 176; LMA, DL/C/266 f. 142, London Consistory Court, Deposition of William Black, 9 June 1729.
66 OBP, September 1775, Matthew Bevan, John Jennings (t17750913–39).
67 Pierce Egan, *Life in London: or, the Day and Night Scenes of Jerry Hawthorn, Esq. and his Elegant Friend Corinthian Tom, Accompanied by Bob Logic, the Oxonian, in their Rambles and Sprees through the Metropolis* (1821), p. 275.
68 LL; 29 Geo. III, c. 49 (1789); George Kearsley, *Kearsley's Annual Eight-penny Tax Tables* (1795), p. 88; 'An Account of the Number and Amount of Licenses granted to Dealers in Horses within the Cities and within the Liberties of London and Westminster, the Parishes of Saint Mary-le-Bone and Saint Pancras, in the County of Middlesex, and in the Borough of Southwark, in the County of Surrey, for the years 1828, 1829 and 1830', in *House of Commons Accounts and Papers*, 8 vols (1830–31), vol. 8: Miscellaneous, p. 176.
69 OBP: December 1824, William Smith (t18241202–93); December 1795, Lockey Hill (t17951202–53).
70 OBP: September 1733, Jonas Pearson (t17330912–72); May 1775, William Howard (t17750531–24).
71 OBP, January 1784, William Saunders (t17840114–65).
72 Oldham, *The Mansfield Manuscripts*, vol. 2, pp. 958–60.
73 Lord Herbert to Messrs. Tattersall, Arlington Street, 19 February 1787, in Herbert (ed.), *Pembroke Papers*, pp. 388–9.
74 [Matthew Bacon], *A New Abridgment of the Law. By a Gentleman of the Middle Temple*, 5 vols (1736–66), vol. 1, p. 52.
75 *Post Man and Historical Account* (19–21 April 1715).
76 Lawrence, *The Horse in All His Varieties*, pp. 153–9.
77 OBP, January 1810, Robert Norton (t18100110–17).
78 [George Stephen], *The Adventures of a Gentleman in Search of a Horse, by Caveat Emptor* (1835), p. 199.
79 Sylas Neville, diary entries for 5, 6, 10 and 18 April 1769, in Cozens-Hardy (ed.), *The Diary of Sylas Neville*, pp. 67–69.

80 Lawrence, *The Horse in All His Varieties*, pp. 146–7.
81 Lawrence, *The Horse in All His Varieties*, pp. 146–7.
82 Henry Bunbury, *An Academy for Grown Horsemen, Containing the Completest Instructions for Walking, Trotting, Cantering, Galloping, Stumbling, and Tumbling* (1787), pp. 4–5. See also Parker, *A View of Society and Manners*, pp. 47–9.
83 Francis Grose, *A Classical Dictionary of the Vulgar Tongue* (1785); James Caulfield, *Blackguardiana: Or, a Dictionary of Rogues, Bawds, Pimps, Whores, Pickpockets, Shoplifters* (1793). See also Lawrence, *The Horse in All His Varieties*, pp. 155–6.
84 William Burdon, *The Gentleman's Pocket-Farrier; Shewing How to Use Your Horse on a Journey. And What Remedies are Proper for Common Misfortunes that May Befal Him on the Road* (1730); Thomas Wallis, *The Farrier's and Horseman's Complete Dictionary* (1767); James Clark, *A Treatise on the Prevention of Diseases Incidental to Horses, from Bad Management in Regard to Stables, Food, Water, Air and Exercise* (Edinburgh, 1788); William Taplin, *The Gentleman's Stable Directory; or, Modern System of Farriery*, 2 vols (1788); Youatt, *The Horse*.
85 Henry Bracken, *Ten Minutes Advice to Every Gentleman Going to Purchase a Horse Out of a Dealer, Jockey, or, Groom's Stables*, first published 1774, reprinted as a prefix to Henry Bracken, *Farriery Improved; or, a Complete Treatise on the Art of Farriery* (12th edn, 1792), pp. 5–6, 14, 20, 22.
86 Youatt, *The Horse*, p. 67.
87 Caulfield, *Blackguardiana*.
88 *Daily Advertiser* (3 March 1778); *World* (21 June 1787 and 23 May 1788); *Morning Post* (2 May 1794).
89 Lawrence, *A Philosophical and Practical Treatise on Horses*, vol. 2, pp. 167–8, and *The Horse in All His Varieties*, p. 147.
90 Anon., *How to Live in London; or, the Metropolitan Microscope, and Stranger's Guide* (1828), pp. 40–1, emphasis in original.
91 OBP: July 1726, John Brakes, Thomas West (t17260711–49); July 1728, William Tilley (t17280717–18); September 1733, Jonas Pearson (t17330912–72); October 1758, Stephen Walles (t17581025–25).
92 Mayhew, *London Labour*, vol. 4, p. 325.
93 OBP: July 1736, Stephen Phillips (t17360721–29); October 1738, John Machell, Richard Wilkinson (t17381011–1); September 1745, James Bridges (t17450911–45); May 1775, William Howard (t17750531–24); October 1780, Richard Brown, James Johnson (t17801018–8).
94 *World* (7 June 1788); OBP: December 1759, William Budd (t17591205–6); September 1748, Thomas Thompson (t17480907–29); January 1771, Samuel Young (t17710116–8); September 1771, Robert Walker (t17710911–78).

95 Mayhew, *London Labour*, vol. 4, p. 325; OBP, October 1770, John Barton (t17701024–37).

96 *World* (20 October 1789 and 27 November 1788); Beattie, *Crime and the Courts*, p. 169.

97 2 & 3 Philip & Mary, c. 7 (1555); 31 Eliz. I, c. 12 (1588); [Stephen], *The Adventures of a Gentleman in Search of a Horse*, p. 218; OBP: October 1801, John Manby, John Manby (t18011028–18); July 1816, John Hardy (t18160710–24); September 1803, Robert Hummerston, Thomas Hummerston (t18030914–95); February 1745, William Joyce (t17450227–18); April 1773, William Parker (t17730421–24).

98 OBP: April 1720, John Ridge (t17200427–23); October 1726, Thomas Hide (t17261012–40); June 1731, Samuel Curlis (t17310602–38); February 1745, William Joyce (t17450227–18); September 1791, John Portsmouth (t17910914–46).

99 OBP: June 1731, Samuel Curlis (t17310602–38); February 1767, William Pattison (t17670218–15).

100 *London Evening Post* (12–14 December 1771); *Morning Post & Daily Advertiser* (12 June 1780); *St James's Chronicle or the British Evening Post* (19–21 June 1788); World (16 May 1789); *St. James's Chronicle or the British Evening Post* (26–29 July 1794); *Sun* (6 June 1800).

101 *Public Advertiser* (6 July 1754); OBP, September 1774, Joseph Doggett (t17740907–50).

102 OBP: September 1778, James Durham (t17780916–1); December 1780, Thomas Brown (t17801206–26); September 1785, John Lloyd (t17850914–103); Orchard, *Tattersalls*, pp. 143, 166.

103 Egan, *Life in London*, pp. 274–6, emphasis in original.

104 Earle, *The Making of the English Middle Class*, pp. 343–44; Barker and Gerhold, *The Rise and Rise of Road Transport*, p. 60.

105 In the 1890s, it was estimated that London housed 40,000 carriage horses alone. See Gordon, *The Horse World of London*, p. 61.

6 Horsing around

1 Longrigg, *The English Squire*, pp. 99–177; P. Borsay, *The English Urban Renaissance: Culture and Society in the Provincial Town, 1660–1770* (Oxford: Oxford University Press, 1989), pp. 178–96.

2 M. Kwint, 'Astley's amphitheatre and the early circus in England, 1768–1830' (PhD thesis, University of Oxford, 1994), pp. 13–80, and 'The circus and nature in late Georgian England', in R. Koshar (ed.), *Histories of Leisure* (Oxford and New York: Berg, 2002), p. 46.

3 *Globe* (30 April 1811); M. Gamer, 'A matter of turf: Romanticism, hippodrama, and legitimate satire', *Nineteenth-Century Contexts*, 28 (2006), 319–21; A. H. Saxon, *Enter Foot and Horse: A History of Hippodrama in England and France* (New Haven: Yale University Press, 1968), p. 90; J. Moody, *Illegitimate Theatre in London, 1770–1840* (Cambridge: Cambridge University Press, 2000), p. 70; Frederick Reynolds, *The Life and Times of Frederick Reynolds*, 2 vols (1826), vol. 2, p. 404.

4 W. C. Oulton, *A History of the Theatres of London… From the Year 1795 to 1817 Inclusive*, 3 vols (1818), vol. 3, p. 119.

5 G. E. Mingay, *English Landed Society in the Eighteenth Century* (London: Routledge and Kegan Paul, 1963), p. 152; Longrigg, *The English Squire*; Donald, *Picturing Animals in Britain*, chs. 6 and 7; L. Colley, *Britons: Forging the Nation 1707–1837* (New Haven and London: Yale University Press, 1992), p. 172; Borsay, *The English Urban Renaissance*, pp. 176–96.

6 S. E. Whyman, *Sociability and Power in Late-Stuart England: The Cultural Worlds of the Verneys 1660–1720* (Oxford: Oxford University Press, 1999); A. Vickery, *Behind Closed Doors: At Home in Georgian England* (New Haven and London: Yale University Press, 2009), pp. 119, 124, 141.

7 Mingay, *English Landed Society*; Longrigg, *The English Squire*.

8 P. Clark, *British Clubs and Societies 1580–1800: The Origins of an Associational World* (Oxford: Oxford University Press, 2001); J. Brewer, *The Pleasures of the Imagination: English Culture in the Eighteenth Century* (London: HarperCollins), pp. 34–50.

9 A. Vickery, *The Gentleman's Daughter: Women's Lives in Georgian England* (New Haven and London: Yale University Press, 1998), pp. 261–4, 272–5; Mingay, *English Landed Society*, pp. 156–7, 205; H. Berry and J. Gregory (eds), *Creating and Consuming Culture in North-East England, 1660–1830* (Aldershot: Ashgate, 2004); Borsay, *The English Urban Renaissance*.

10 Henry Fielding, *The History of the Adventures of Joseph Andrews, and his Friend Mr Abraham Adams*, 2 vols (1742), vol. 2, pp. 103–8, and *The History of Tom Jones, a Foundling*, 4 vols (1749), vol. 3, pp. 138–9.

11 James Thompson, *The Seasons* (1726–30, revised 1744); John Aldington, *A Poem on the Cruelty of Shooting* (1769); Oliver Goldsmith, *The Deserted Village, A Poem* (1770); Edward Lovibond, 'On Rural Sports', in Edward Lovibond, *Poems on Several Occasions* (1785), pp. 47–53; William Cowper, *The Task* (1785), Book 1: The Sofa, Book 3: The Garden and Book 4: The Winter Evening; D. Landry, *The Invention of the Countryside: Hunting, Walking, and Ecology in English Literature, 1671–1831* (Basingstoke: Palgrave Macmillan, 2001), p. 115;

Porter, 'Enlightenment London and urbanity', pp. 27–9 and 'The urban and the rustic in Enlightenment London', pp. 176–94.

12 S. Brown, 'The human–animal bond and self psychology: Toward a new understanding', *Society and Animals*, 12 (2004), 70, 79; A. M. Beck and A. H. Katcher, *Between Pets and People: The Importance of Animal Companionship* (West Lafayette, IN: Purdue University Press, 1996); E. S. Wolf, *Treating the Self: Elements of Clinical Self Psychology* (New York: Guilford Press, 1988), p. 63.

13 K. Kete, *The Beast in the Boudoir: Petkeeping in Nineteenth-Century Paris* (Berkeley and Los Angeles: University of California Press, 1994); Ritvo, *The Animal Estate*, ch. 2; Tague, *Animal Companions*.

14 Trusler, *The London Advisor and Guide*, pp. 120–1.

15 Vickery, *Behind Closed Doors*, pp. 12–13, 124; Whyman, *Sociability and Power*, pp. 93–107; Landry, *The Invention of the Countryside*, pp. 163–6.

16 Joseph Browne, *The Circus: Or, British Olympicks. A Satyr on the Ring in Hyde-Park* (1709); *Spectator*, no. 88 (11 June 1711); Henri Misson, *M. Misson's Memoirs and Observations in His Travels Over England*, trans. John Ozell (1719), p. 126; Thomas Salmon, *Modern History: Or, the Present State of All Nations*, 26 vols (1725–35), vol. 15, p. 339; *The London Spy Revived* (6 December 1736); Walter Thornbury, *Old and New London*, 6 vols (1878), vol. 4, pp. 375–80; N. Braybrooke, *London Green: The Story of Kensington Gardens, Hyde Park, Green Park and St. James's Park* (London: Victor Gollancz, 1959), pp. 60, 74.

17 S. Lasdun, *The English Park: Royal, Private, and Public* (New York: Vendome Press, 1992), pp. 76, 124; Ann Thicknesse, *A Letter from Miss F—d, Addressed to a Person of Distinction* (1761), p. 6; Braybrooke, *London Green*, p. 63.

18 *Argus* (11 April 1791), emphasis in the original.

19 Richard Hodgkinson, journal entry for 2 March 1794, in Wood and Wood (eds), *A Lancashire Gentleman*, p. 41.

20 *Public Advertiser* (29 February 1792).

21 Henry Angelo, *Reminiscences of Henry Angelo with Memoirs of His Late Father and Friends*, 2 vols (1830), vol. 1, pp. 35–6; Richard Berenger, *The History and Art of Horsemanship*, 2 vols (1771); John Adams, *Analysis of Horsemanship: Teaching the Whole Art of Riding in the Manege, Military, Hunting, Racing, or Travelling System* (Edinburgh, 1799), vol. 1, preface; Worsley, *The British Stable*, ch. 7; D. Landry, 'Learning to ride in early modern Britain, or, the making of the English Hunting Seat', in K. Raber and T. J. Tucker (eds), *The Culture of the Horse. Status, Discipline, and Identity in the Early Modern World* (Basingstoke: Palgrave Macmillan, 2005), p. 331.

22 Carter's: *World & Fashionable Advertiser* (9 October 1787 and 13 December 1787); Pantheon school: *Morning Herald* (25 March 1799); Angelo's Academy: *Lloyds Evening Post* (21 November 1764); Park & Son: *Morning Chronicle* (20 December 1787). For free entry, see Carter's: *Morning Post & Daily Advertiser* (4 February 1780) and Wright's at Gray's Inn Lane Road: *World* (9 September 1791).
23 Johnson's in Upper Moorfields allowed gentlemen keeping horses at livery to 'practice at the bar, and benefit their health' gratis: *Morning Post & Daily Advertiser* (6 November 1786). For Carter's subscription offer, see *Morning Herald & Daily Advertiser* (30 November 1780). Carter's charged 5s for renting a horse to ride in the park and 3s 6d for use in his riding house: *Morning Post & Daily Advertiser* (30 October 1778). Wright's let out saddle horses for ladies and gentlemen: *World* (27 April 1792).
24 Carter's: *Morning Herald & Daily Advertiser* (14 February 1783); Astley's: *Public Advertiser* (16 January 1779).
25 Pantheon school: *Morning Post and Gazetteer* (25 December 1798); Park & Son: *Morning Chronicle* (20 December 1787); Royal Circus: *Morning Post* (18 April 1800); Astley's: *Gazetteer* (27 July 1779).
26 Philip Dormer Stanhope, Earl of Chesterfield to Philip Stanhope, 16 May 1751, in D. Roberts (ed.), *Lord Chesterfield's Letters* (Oxford: Oxford University Press, 1998), p. 230; *Lloyd's Evening Post* (21 November 1764); *Gazetteer* (4 March 1767); *London Evening Post* (7 May 1771); *Morning Post & Daily Advertiser* (10 January 1778); Sylas Neville, diary entry for 26 February 1768, in Cozens-Hardy (ed.), *The Diary of Sylas Neville*, p. 30; Brewer, *The Pleasures of the Imagination*, pp. 107, 547; J. H. Plumb, 'The new world of children in eighteenth-century England', *Past and Present*, 67 (1975), 72, 79.
27 *Morning Post & Gazetteer* (25 December 1798); *Morning Post* (4 February 1780); *Morning Post* (31 May 1830).
28 Lady Mary Coke, diary entries, November 1766 and May 1767, in J. A. Home (ed.), *The Letters and Journals of Lady Mary Coke*, 4 vols (Edinburgh, 1889–96), vol. 1, p. 247; *World & Fashionable Advertiser* (9 October 1787 and 13 December 1787); Angelo, *Reminiscences of Henry Angelo*, vol. 1, p. 6; *Morning Post & Fashionable World* (3 July 1795); Frances Anne Kemble, diary entries, June 1831 and 13 December 1831, in Frances Anne Kemble, *Records of a Girlhood*, 3 vols (New York, 1879), pp. 605.
29 Colley, *Britons*, p. 172.
30 P. Gauci, *Emporium of the World: The Merchants of London 1660–1800* (London: Hambledon Continuum, 2007), p. 75.

31 LMA, ACC/1017/0944, Diary of John Eliot (III), 1757; TNA, J90/12–14, Diary of Thomas Bridge, 1760–1811, entries for 8 January, 5 April and 10 April 1762.

32 TNA, J90/12–14, Diary of Thomas Bridge, diary entry for 10 April 1762; LMA, ACC/1017/0944, Diary of John Eliot; Sylas Neville, diary entry for 11 February 1768, in Cozens-Hardy (ed.), *The Diary of Sylas Neville*, p. 30; H. Greig, '"All together and all distinct": Public sociability and social exclusivity in London's pleasure gardens, *c.* 1740–1800', *Journal of British Studies*, 51 (2012), 50–75; P. Gauci, *The Politics of Trade: The Overseas Merchant in State and Society, 1660–1720* (Oxford: Oxford University Press, 2001), p. 88.

33 [William Cowper], 'The Entertaining and Facetious History of John Gilpin…', *The Public Advertiser*, 14 November 1782; William Cowper, 'The Diverting History of John Gilpin, Shewing How He Went Farther Than He Intended, and Came Safe Home Again', in Cowper, *The Task*, pp. 342–59.

34 Anon., *Letters from an Irish Student in England to his Father in Ireland*, 2 vols (1809), vol. 1, pp. 195–6.

35 Defoe, *A Tour* (1986), pp. 48–9, 167–8; TNA, J90/12–14, Diary of Thomas Bridge; LMA, ACC/1017/0944, Diary of John Eliot.

36 Borsay, *The English Urban Renaissance*, pp. 182, 302, Appendix 7; 13 Geo. II, c. 19 (1740); *Whitehall Evening Post* (29–31 August 1749); *Public Advertiser* (26 September 1753); LMA, MJ/SP/1755/09, Middlesex Sessions Papers, 1755, Justices' working documents, September 1755 (LL, LMSMPS504440009).

37 Borsay, *The English Urban Renaissance*, p. 185; *Gazetteer & New Daily Advertiser* (2 September 1771).

38 Anon., 'Invitation to Epsom-Races. A Song. By Philo-Bumper', in Anon., *Eclipse Races (Addressed to the Ladies:) Being an Impartial Account of the Celestial Coursers and their Riders… By Philo-Pegasus, a Lover of Truth* (1764), p. 20.

39 *Whitehall Evening Post* (5–7 May 1785); *Evening Mail* (25–28 May 1792).

40 John Cook, *Observations on Fox Hunting, and the Management of Hounds in the Kennel and the Field* (1826), pp. 148–9, emphasis in original.

41 Henry Playford, *Wit and Mirth: Or, Pills to Purge Melancholy: Being a Collection of the Best Merry Ballads and Songs, Old and New*, 4 vols (1707), vol. 2, p. 43.

42 Thomas Rowlandson, *Easter Monday, or the Cockney Hunt* (hand-coloured etching, 1811).

43 Robert Smith Surtees, *Jorrocks's Jaunts and Jollities: The Hunting, Shooting, Racing, Driving, Sailing, Eating, Eccentric and Extravagant Exploits of that*

Renowned Sporting Citizen Mr. John Jorrocks of St. Botolph Lane and Great Coram Street (2nd edn, 1843), pp. 21, 56–69; J. Welcome, *The Sporting World of R. S. Surtees* (Oxford: Oxford University Press, 1982), p. 63.

44 *The Sporting Magazine*, vol. 1, 2nd series, no. 1 (1830), p. 413.

45 Cook, *Observations on Fox Hunting*, pp. 148–9.

46 R. Longrigg, *The History of Foxhunting* (Basingstoke: Macmillan, 1975), p. 81; H. R. Taylor, *The Old Surrey Fox Hounds: A History of the Hunt From its Earliest Days to the Present Time* (London: Longmans, Green & Co., 1906); *St James's Chronicle* (14 January 1775).

47 *Sun* (16 August 1796).

48 Cobbett, *Rural Rides*, pp. 259–60.

49 *World* (28 February 1788); Surtees, *Jorrocks's Jaunts*, pp. 34–8, 42–7, 54. See also *Illustrated London News* (6 April 1867).

50 Surtees, *Jorrocks's Jaunts*, p. 8; OBP, February 1782, Edward Wilkins (t17820220–2).

51 *World* (1 August 1791).

52 Worsley, *The British Stable*, pp. 168–9; Lady Mary Coke, diary entry, May 1767, in Home (ed.), *The Letters and Journals*, vol. 1, p. 247; Frances Anne Kemble, diary entry, 19 December 1831, in Kemble, *Records of a Girlhood*, p. 605.

53 *Morning Herald & Daily Advertiser* (30 November 1780); *Morning Post & Daily Advertiser* (6 November 1786); *Star* (19 May 1795).

54 TNA, J90/12–14, Diary of Thomas Bridge, 1762.

55 Lady Mary Coke, diary entry, February 1769, in Home (ed.), *The Letters and Journals*, vol. 3, p. 495; Landry, *The Invention of the Countryside*, p. 166; *London Evening Post* (22–25 September 1753); *Independent Chronicle* (6–9 October 1769).

56 Sylas Neville, diary entries for 6 April and 18 May 1769, in Cozens-Hardy (ed.), *The Diary of Sylas Neville*, pp. 67, 69.

57 L. E. Klein, 'Politeness for plebes: Consumption and social identity in early eighteenth-century England', in A. Bermingham and J. Brewer (eds), *The Consumption of Culture, 1600–1800: Image, Object, Text* (London: Routledge, 1995), p. 367; Greig, '"All together and all distinct"', p. 51; See also Borsay, *The English Urban Renaissance*, pp. 277–9; R. Porter, *English Society in the Eighteenth Century* (2nd edn, Harmondsworth: Penguin, 1990), p. 232; Langford, *A Polite and Commercial People*, p. 102.

58 Lady Mary Coke, diary entry, May 1767, in Home (ed.), *The Letters and Journals*, vol. 1, p. 247.

59 *London Evening Post* (9 June 1763).

60 Lady Mary Coke, diary entry, May 1767, in Home (ed.), *The Letters and Journals,* vol. 1, p. 247.

61 P. Carter, *Men and the Emergence of Polite Society, Britain 1660–1800* (Harlow: Longman, 2001); M. Cohen, 'Manliness, effeminacy and the French: Gender and the construction of national character in eighteenth-century England', in T. Hitchcock and M. Cohen (eds), *English Masculinities, 1660–1800* (Harlow: Longman, 1999), p. 51.

62 Colley, *Britons*, p. 172.

63 Lawrence, *A Philosophical and Practical Treatise on Horses*, vol. 1, p. 248.

64 James Boswell, journal entry, 21 August 1773, in Boswell, *The Journal of a Tour to the Hebrides*, p. 201; D. Wahrman, *The Making of the Modern Self: Identity and Culture in Eighteenth-Century England* (New Haven and London: Yale University Press, 2004).

65 Anon., *Horse and Away to St. James's Park Or, a Trip for the Noontide Air Who Rides Fastest, Miss Kitty Fisher or her Gay Gallant* (printed at Strawberry Hill, 1760); M. Pointon, 'The lives of Kitty Fisher', *British Journal for Eighteenth-Century Studies*, 27 (2004), 77–97.

66 See, for instance: Mr. Oakman, song 186, in *The Busy Bee, or, Vocal Repository. Being a Selection of the Most Favourite Songs, &c. Contained in the English Operas, That Have Been Sung at the Public Gardens, and Written for Select Societies…*, 3 vols (1790), vol. 2, p. 214; [Charles Dunster], *St. James's Street, A Poem, in Blank Verse. By Marmaduke Milton, Esq.* (1790), p. 21; Charlotte Turner Smith, *Marchmont: A Novel*, 4 vols (1796), vol. 1, pp. 139–40; [William Beckford], *Modern Novel Writing, or the Elegant Enthusiast; and Interesting Emotions of Arabella Bloomville*, 2 vols (1796), vol. 1, pp. 51–3.

67 J. Newman, '"An insurrection of loyalty": The London volunteer regiments' response to the invasion threat', in M. Philp (ed.), *Resisting Napoleon: The British Response to the Threat of Invasion, 1797–1815* (Aldershot: Ashgate, 2006), pp. 75–90.

68 Thomas Rowlandson, *The Light Horse Volunteers of London & Westminster, Commanded by Coll Herries, Reviewed by His Majesty on Wimbledon Common, 5th July 1798* (etching and aquatint, 1798).

69 *Observer* (22 April 1798); Colley, *Britons*, pp. 152–3, 184–92.

70 B. Heller, 'Leisure and pleasure in London society, 1760–1820: An agent-centred account' (PhD thesis, University of Oxford, 2009), pp. 220–4.

71 Letter from Frances Anne Kemble to H [Harriet St. Leger], 17 January 1830, in Kemble, *Records of a Girlhood*, p. 605.

72 Frances Anne Kemble, diary entries, 19 December 1831 and January 1832, in Kemble, *Records of a Girlhood*, p. 605.

73 G. R. Williams, *London in the Country: The Growth of Suburbia* (London: Hamish Hamilton, 1975), pp. 1, 5.

74 LMA, ACC/1017/0944, Diary of John Eliot; TNA, J90/12–14, Diary of Thomas Bridge, 1772; *The Spectator*, no. 412 (23 June 1712).

75 G. Cheyne, *An Essay of Health and Long Life* (1724), p. 120; John Wesley, *Primitive Physic: Or, An Easy and Natural Method of Curing Most Diseases* (24th edn, 1792), pp. xii–xiii; John Wesley, *The Journal of the Rev. John Wesley, A. M.*, 4 vols (1827), vol. 4, p. 77; D. Madden, *'A Cheap, Safe and Natural Medicine': Religion, Medicine and Culture in John Wesley's Primitive Physic* (Amsterdam: Rodopi, 2007), pp. 3, 185.

76 John Allen, diary entries, 27, 29 April and 31 May 1777, in Sturge (ed.), *Leaves from the Past*, pp. 45, 58.

77 Adams, *Analysis of Horsemanship*, pp. xiii–xiv.

78 P. Shepard, *The Others: How Animals Made Us Human* (Washington, DC: Island Press, 1996), p. 251; L. Birke, 'Talking about horses: Control and freedom in the world of "natural horsemanship"', *Society and Animals*, 16 (2008), 115. See also A. H. Kidd and R. M. Kidd, 'Seeking a theory of the human/companion animal bond', *Anthrozoös*, 1 (1987), 140–57; J. Serpell, *In the Company of Animals: A Study of Human–Animal Relationships* (1986; Cambridge: Cambridge University Press, 1996); C. C. Wilson and D. C. Turner (eds), *Companion Animals in Human Health* (Thousand Oakes, CA: Sage Publications, 1998).

79 Raber and Tucker (eds), *The Culture of the Horse*; Landry, *Noble Brutes*, ch. 2; Birke, 'Talking about horses', 118.

80 William Bridges Adams, *English Pleasure Carriages: Their Origin, History, Varieties, Materials, Construction, Defects, Improvements, and Capabilities* (1837), pp. 198–9.

81 Sylas Neville, diary entries for 5, 6, 10, 13, 18 April, 18 May 1769, 13 October 1772 and 8 October 1784, in Cozens-Hardy (ed.), *The Diary of Sylas Neville*, pp. 67–9, 182, 323.

82 Lady Mary Coke, diary entry, 25 October 1768, in Home (ed.), *The Letters and Journals*, vol. 2, p. 440; TNA, J90/12–14, Diary of Thomas Bridge, 1762.

83 Borsay, *The English Urban Renaissance*, p. 181.

84 Thomas Rowlandson, *A Crowded Race Meeting*, watercolour with pen in brown and grey ink on paper, *c.* 1805–1810, Yale Center for British Art, Paul Mellon Collection.

85 Grosley, *A Tour to London*, vol. 2, p. 158.

86 *London Spy Revived* (13 September 1736).

87 *Gazetteer & New Daily Advertiser* (2 September 1771).

88 D. Masson (ed.), *The Collected Writings of Thomas de Quincey*, 14 vols (Edinburgh, 1889–90), vol. 13, pp. 283–4.

7 Watchdogs

1 OBP: April 1744, James Cooper (t17440404–29); July 1765, Mary Wood (t17650710–57); April 1819, Patrick Hagan (t18190421–205).
2 *Public Advertiser* (16 September 1775).
3 Cockayne, *Hubbub*; P. Howell, *At Home and Astray: The Domestic Dog in Victorian Britain* (Charlottesville: University of Virginia Press, 2015); Tague, *Animal Companions*.
4 Greig, '"All together and all distinct"', 50–75; S. Lloyd, 'Ticketing the British eighteenth century: "A thing… never heard of before"', *Journal of Social History*, 46 (2013), 843–71; Meldrum, *Domestic Service and Gender*, ch. 4; A. Vickery, 'An Englishman's home is his castle? Thresholds, boundaries and privacies in the eighteenth-century London house', *Past and Present*, 199 (2008), 159–62; T. Hitchcock and R. Shoemaker, *London Lives: Poverty, Crime and the Making of a Modern City, 1690–1800* (Cambridge: Cambridge University Press, 2015).
5 C. Pearson, 'Beyond "resistance": Rethinking nonhuman agency for a "more-than-human" world', *European Review of History*, 22 (2015), 709–25, and 'Canines and contraband: Dogs, nonhuman agency and the making of the Franco-Belgian border during the French Third Republic', *Journal of Historical Geography*, 54 (2016), 50–62.
6 J. M. Beattie, *Policing and Punishment in London, 1660–1750: Urban Crime and the Limits of Terror* (Oxford: Oxford University Press, 2001) and *The First English Detectives: The Bow Street Runners and the Policing of London, 1750–1840* (Oxford: Oxford University Press, 2012); E. A. Reynolds, *Before the Bobbies: The Night Watch and Police Reform in Metropolitan London, 1720–1830* (Basingstoke: Macmillan, 1998).
7 Massie, *An Essay on the Many Advantages*, p. 23. See also Hanway, *A Letter to Mr. John Spranger*, p. 36.
8 By 1737, almost 700 watchmen worked in the City of London; Westminster and the urban parishes of Middlesex employed a similar number; see Hitchcock, *Down and Out*, pp. 153–6; Beattie, *Policing and Punishment*, p. 195.
9 Patrick Colquhoun, *A Treatise on the Police of the Metropolis: Containing a Detail of the Various Crimes and Misdemeanours by which Public and Private Property and Security are, at Present, Injured and Endangered: and Suggesting Remedies for Their Prevention* (4th edn, 1797), pp. 98–9.

10 L. Radzinowicz, *A History of the English Criminal Law and its Administration from 1750*, 5 vols (London: Stevens & Sons, 1948–86), vol. 2, p. 372.
11 Vickery, *Behind Closed Doors*, pp. 31, 38–9.
12 *Common Sense or The Englishman's Journal* (10 March 1739).
13 S. Tanner, T. Muñoz and P. Hemy Ros, 'Measuring mass text digitization quality and usefulness: Lessons learned from assessing the OCR accuracy of the British Library's 19th century online newspaper archive', *D-Lib Magazine*, 15 (2009), available at www.dlib.org/dlib/july09/munoz/07munoz.html (accessed 14 January 2019).
14 *Daily Journal* (14 April 1730).
15 E. Snell, 'Discourses of criminality in the eighteenth-century press: The presentation of crime in *The Kentish Post*, 1717–1768', *Continuity and Change*, 22 (2007), 16, 25, 27.
16 Reynolds, *Before the Bobbies*, p. 30; Beattie, *Policing and Punishment*, p. 42; Andrew, *Philanthropy and Police*, p. 164.
17 P. King, 'Newspaper reporting and attitudes to crime and justice in late-eighteenth- and early-nineteenth-century London', *Continuity and Change*, 22 (2007), 76; R. Shoemaker, 'Worrying about crime: Experience, moral panics and public opinion in London, 1660–1800', *Past and Present*, 234 (2017), 71–100.
18 *Gazetteer & New Daily Advertiser* (16 May 1764).
19 *Gazetteer & New Daily Advertiser* (21 September 1765).
20 *London Pack or New Lloyd's Evening Post* (30 October–1 November 1797).
21 King, 'Newspaper reporting and attitudes to crime and justice', 73–112.
22 W. S. Lewis (ed.), *The Yale Edition of Horace Walpole's Correspondence*, 48 vols (New Haven: Yale University Press, 1937–83), pp. xxxii, 167.
23 OBP: October 1731, Francis Richmond (t17311013–31); December 1678 (16781211); May 1698, William Parr, John Ingram (t16980504–49); January 1698, Grace Turner (t16980114–45); July 1698, John Davenport (t16980720–3).
24 OBP, June 1829, James Butler (t18290611–318). For an example of a dog raising the alarm in Victorian London, see OBP, March 1875, John Phillips (t18750301–200).
25 J. H. Langbein, *The Origins of Adversary Criminal Trial* (Oxford: Oxford University Press, 2003), p. 185; R. Shoemaker, 'The Old Bailey Proceedings and the representation of crime and criminal justice in eighteenth-century London', *Journal of British Studies*, 47 (2008), 559–80.
26 Shoemaker, 'Worrying about crime', 100.
27 OBP, February 1827, John Duxberry, William Fox, Rebecca Mullins (t18270215–57).

28 G. J. Adams and K. G. Johnson, 'Guard dogs: Sleep, work and the behavioural responses to people and other stimuli', *Applied Animal Behaviour Science*, 46 (1995), 113–14.
29 OBP, October 1818, Stephen Morris (t18181028–80).
30 *Whitehall Evening Post* (7–9 October 1784).
31 C. Pearson, 'Dogs, history, and agency', *History and Theory*, 52 (2013), 136.
32 Oliver Goldsmith, *An History of the Earth, and Animated Nature*, 8 vols (1774), vol. 3, pp. 251–3.
33 [Edward Augustus Kendall], *Keeper's Travels in Search of His Master* (1798), pp. 65–67; Ebenezer Sibly, *Magazine of Natural History. Comprehending the Whole Science of Animals, Plants, and Minerals*, 14 vols (1794–1807), vol. 2, pp. 4–5.
34 *Morning Post & Daily Advertiser* (14 September 1775).
35 *Gazetteer & New Daily Advertiser* (27 November 1767).
36 G. J. Adams, 'Sleep-wake cycles and other night-time behaviours of the domestic dog *Canis familiaris*', *Applied Animal Behaviour Science*, 36 (1993), 233–48.
37 A. R. Ekirch, 'Sleep we have lost: Pre-industrial slumber in the British Isles', *American Historical Review*, 106 (2001), 344, 368–9.
38 Blackstone, *Commentaries on the Laws of England*, vol. 4, p. 149.
39 Goldsmith, *An History of the Earth*, vol. 3, p. 253.
40 [John Lawrence], *The Sportsman's Repository; Comprising a Series of Highly-Finished Engravings, Representing the Horse and the Dog, in All Their Varieties; Executed in the Line Manner, by John Scott, from Original Paintings by Marshall, Reinagle, Gilpin, Stubbs, and Cooper…* (1820), p. 160.
41 [Lawrence], *The Sportsman's Repository*, p. 161.
42 Tague, *Animal Companions*, p. 94. See also M. Ellis, 'Suffering things: Lapdogs, slaves, and counter-sensibility', in M. Blackwell (ed.), *The Secret Life of Things: Animals, Objects, and It-Narratives in Eighteenth-Century England* (Cranbury, NJ: Associated University Presses, 2007), pp. 101–2.
43 Goldsmith, *An History of the Earth*, vol. 3, pp. 251, 264, 268.
44 Tague, *Animal Companions*, pp. 4–6, 228.
45 *St James' Chronicle* (15–18 November 1777); *Public Advertiser* (18 June 1783).
46 OBP, September 1795, Elizabeth Spiers, Sarah Rote (t17950916–30).
47 *London Chronicle* (27–29 April 1784).
48 Goldsmith, *An History of the Earth*, vol. 3, p. 255; *Public Advertiser* (16 September 1775). See also *London Daily Advertiser* (16 November 1751).
49 *World* (22 March 1787). See also OBP, May 1814, Rebecca Eveleigh (t18140525–2).

50 OBP: July 1778, John Holt, John Davis, Andrew Carleton, Alexander Carleton (t17780715–88); December 1784, William Astill, John Ellis (t17841208–10); September 1825, Edward Mason, William Crook, Ann Gable, Samuel Crook (t18250915–344).
51 *Morning Chronicle & London Advertiser* (30 May 1776).
52 Tague, *Animal Companions*, p. 172.
53 M. Makepeace, *The East India Company's London Workers: Management of the Warehouse Labourers, 1800–1858* (Woodbridge: The Boydell Press, 2010), p. 11.
54 *Gazetteer & New Daily Advertiser* (14 March 1769).
55 Adams and Johnson, 'Guard dogs', 111.
56 S. Coren, *How Dogs Think: Understanding the Canine Mind* (London: Pocket, 2005), p. 166, emphasis in the original.
57 William Ellis, *A Compleat System of Experienced Improvements, made on Sheep, Grass-Lambs, and House-Lambs: Or, The Country Gentleman's, the Grasier's, the Sheep-Dealer's, and The Shepherd's Sure Guide* (1749), p. 20.
58 Samuel Pepys, diary entry, 15 January 1660, in R. Lathan and W. Matthews (eds), *The Diary of Samuel Pepys: A New and Complete Transcription*, 11 vols (London: HarperCollins, 2000), vol. 1, pp. 17–18.
59 *Whitehall Evening Post* (4–6 December 1798).
60 [Lawrence], *The Sportsman's Repository*, p. 160.
61 S. Yin, 'A new perspective on barking in dogs (*Canis familiaris*)', *Journal of Comparative Psychology*, 116 (2002), 189–93.
62 OBP, January 1800, John Moore (t18000115–2).
63 OBP, May 1814, Rebecca Eveleigh (t18140525–2).
64 *St James' Chronicle* (21–24 May 1774).
65 OBP, December 1784, William Astill, John Ellis (t17841208–10). See also, OBP, April 1827, Thomas Shoes Smith, Edward James Plowman (t18270405–231).
66 OBP, July 1786, Alexander Bell, John Strong (t17860719–43).
67 *St James' Chronicle* (21–24 May 1774); *Whitehall Evening Post* (7–9 October 1784). See also Goldsmith, *An History of the Earth*, vol. 3, pp. 251–6.
68 *London Daily Advertiser* (16 November 1751).
69 Goldsmith, *An History of the Earth*, vol. 3, pp. 253–7.
70 [Lawrence], *The Sportsman's Repository*, p. 161.
71 S. M. Kundey, A. De Los Reyes, E. Royer, S. Molina, *et al.*, 'Reputation-like inference in domestic dogs (*Canis familiaris*)', *Animal Cognition*, 14 (2011), 291–302.
72 *Middlesex Journal & Evening Advertiser* (1–3 February 1774).

73 OBP, December 1769, John Randal (t17691206–16).

74 OBP, September 1825, Edward Mason, William Crook, Ann Gable, Samuel Crook (t18250915–344).

75 OBP, June 1829, James Butler (t18290611–318).

76 Af. 23–23.2; Benjamin Thorpe, *Ancient Laws and Institutes of England*, 2 vols (1840), vol. 1 p. 79; W. Rastell, *Registrum Omnium Brevium Tam Originalium Quam Iudicalium* (1531), ff. 110v-111r; OBP, September 1684, Thomas Jeffes (t16840903–23); Zachary Babington, *Advice to Jurors in Cases of Blood* (1677), pp. 141–3; Matthew Hale, *Historia Placitorum Coronæ. The History of the Pleas of the Crown*, 2 vols (1736), vol. 1, p. 431; Richard Burn, *The Justice of the Peace, and Parish Officer*, 2 vols (1755), vol. 2, p. 8; Blackstone, *Commentaries on the Laws of England*, vol. 4, p. 130; R. C. Palmer, *English Law in the Age of the Black Death, 1348–1381: A Transformation of Governance and Law* (Chapel Hill: University of North Carolina Press, 1993), pp. 238–40; E. Cockayne, 'Who did let the dogs out? Nuisance dogs in late-medieval and early modern England', in L. D. Gelfand (ed.), *Our Dogs, Our Selves: Dogs in Medieval and Early Modern Art, Literature, and Society* (Leiden: Brill, 2016), pp. 65–6.

77 Blackstone, *Commentaries on the Laws of England*, vol. 4, p. 130; William Oldnall Russell and Daniel Davis, *A Treatise on Crimes and Misdemeanors*, 2 vols (Boston, 1824), vol. 1, pp. 438–9.

78 Isaac Éspinasse, *Reports of Cases Argued and Ruled at Nisi Prius, in the Courts of King's Bench, and Common Pleas, from Easter Term 33 George III 1793 to Hilary Term 36 George III, 1796* (Dublin, 1797), pp. 203–4.

79 *Sun* (2 August 1798).

80 Isaac Cruikshank, *Effects of the Dog Tax* (hand-coloured etching, 1796).

81 *The Gentleman's Magazine*, 30 (1760), pp. 353–4, 371–3, 392, 464–5.

82 The *Parliamentary History of England, from the Earliest Period to the Year 1803*, vol. 32 (1818), 5 April 1796, col. 995; I. H. Tague, 'Eighteenth-century English debates on a dog tax', *The Historical Journal*, 51 (2008), 901–20.

83 Edward Barry, *On the Necessity of Adopting Some Measures to Reduce the Present Number of Dogs; with a Short Account of Hydrophobia, and the Most Approved Remedies Against it, a Letter, to Frances Annesley, Esq., MP for the Borough of Reading, and one of the Trustees of the British Museum* (Reading, 1796), p. 6; George Morland, *The Miseries of Idleness* (oil on canvas, 1790), Scottish National Gallery.

84 William King, *Reasons and Proposals for Laying a Tax upon Dogs. Humbly Addressed to the Honourable House of Commons* (Reading, 1740), pp. 13–16.

85 *The Parliamentary Register; or History of the Proceedings and Debates of the House of Commons*, vol. 44 (1796), p. 516.

86 *The Parliamentary Register*, vol. 44, p. 506.

87 As reported in the *Sun* newspaper (26 April 1796). See also *The Parliamentary Register*, vol. 44, p. 508.

88 Anon., *The Lamentation of a Dog, on the Tax, and its Consequences. Addressed to the Right Hon. William Pitt. With Notes by Scriblerus Secundus* (1796).

89 36 Geo. III, c. 124 (1796).

90 Westminster City Archives, F2505, St Martin-in-the-Fields parish records, Schedules of the sums assessed on the several wards in respect of duties upon shops, servants, horses, carriages and wagons, with the names of defaulters, 1785–98; and E3159B, St Margaret Westminster parish records, Duties on servants, carriages, horses and dog tax, commissioners' minute book, 1798–1804; Thornbury, *Old and New London*, vol. 3, pp. 149–60.

Conclusion

1 Berger 'Why look at animals?', p. 12.

2 Hammond, 'The Industrial Revolution and discontent'.

3 *London Chronicle* (18 October 1820).

4 Donald, '"Beastly sights"', 514–16; Kean, *Animal Rights*, p. 13; Ritvo, *The Animal Estate*, pp. 125–6.

5 Boddice, *A History of Attitudes and Behaviours Toward Animals*, pp. 29, 344; E. Griffin, *Blood Sport: Hunting in Britain Since 1066* (New Haven and London: Yale University Press, 2007), pp. 142–9. See also R. Preece, 'Thoughts out of season on the history of animal ethics', *Society and Animals*, 15 (2007), 365, 368, 370–1.

6 LMA, CLA/016/AD/02/006, Second report.

7 3 Geo. IV, c. 71 (1822); Kean, *Animal Rights*, pp. 31–6; Donald, '"Beastly sights"', 514–44, and *Picturing Animals*, pp. 224–5; K. Shevelow, *For the Love of Animals: The Rise of the Animal Protection Movement* (New York: Henry Holt & Co., 2008).

8 LMA, CLA/016/AD/02/006, Second report, p. 20.

9 LMA, CLA/016/AD/02/006, Second report, pp. 19–21, 34–5, 55.

10 Samuel Johnson, *A Dictionary of the English Language*, 2 vols (2nd edn, 1755–56).

11 F. M. L. Thompson, 'Nineteenth-century horse sense', *EcHR*, 29 (1976), 60–81.

12 Best Foot Forward Ltd., *City Limits: A Resource Flow and Ecological Footprint Analysis of Greater London*, commissioned by the Chartered Institution of Wastes Management, Northampton, 2002, available at www.citylimitslondon.com/downloads/Complete%20report.pdf (accessed 14

January 2019), p. 12; J. Meikle, 'Pigs in the city', *Guardian* (11 August 1999); C. Cadwalladr, 'Urban farms: Can you source a complete meal from inside the M25?', www.theguardian.com/lifeandstyle/2010/jun/20/urban-farms-local-food (published 20 June 2010; accessed 14 January 2019); P. J. Atkins, 'Is it urban? The relationship between food production and urban space in Britain, 1800–1950', in M. Hietala and T. Vahtikari (eds), *The Landscape of Food: The Food Relationship of Town and Country in Modern Times* (Helsinki: Finnish Literature Society, 2003), pp. 133, 139.

13 This research was conducted by Dairy Farmers of Britain as part of its Grass is Greener educational campaign and involved a poll of more than 1,000 children aged eight to fifteen; *Daily Telegraph*, 'Cows lay eggs and bacon comes from sheep, say children in farmers' poll', www.telegraph.co.uk/news/uknews/1544049/Cows-lay-eggs-and-bacon-comes-from-sheep-say-children-in-farmers-poll.html (published 28 February 2007; accessed 14 January 2019).

14 D. B. Freeman, *A City of Farmers: Informal Urban Agriculture in the Open Spaces of Nairobi, Kenya* (Montreal: McGill-Queen's University Press, 1991); B. Mbiba, *Urban Agriculture in Zimbabwe: Implications for Urban Management and Poverty* (Aldershot: Avebury, 1995); G. Stanhill, 'An urban agro-ecosystem: The example of nineteenth-century Paris', *Agro-Ecosystems*, 3 (1977), 269–84.

15 K. Helmore and A. Ratta, 'The surprising yields of urban agriculture', *Choices* (UNDP), 4 (1995), 22–7; Atkins, 'Is it urban?', pp. 133–5, 139.

16 Steinfeld et al., *Livestock's Long Shadow*, pp. 230, xxi. See also F. Fuller, F. Tuan and E. Wailes, 'Rising demand for meat: Who will feed China's hogs?', in F. Gale (ed.), *China's Food and Agriculture: Issues for the 21st Century*, Agriculture Information Bulletin, 775 (Washington, DC: US Department of Agriculture, 2002), pp. 17–19.

17 A. Viljoen and K. Bohn (eds), *Second Nature Urban Agriculture: Designing Productive Cities* (London: Routledge, 2014); C-T. Soulard, C. Perrin and E. Valette (eds), *Toward Sustainable Relations Between Agriculture and the City* (Cham: Springer International Publishing, 2017); A. M. G. A. WinklerPrins (ed.) *Global Urban Agriculture* (Wallingford: CABI, 2017).

Select bibliography

For secondary literature, see the references in the main text.

Manuscripts

British Library, London

Add MS 27828: Place Papers, vol. 40, 'Manners and Morals', 4, fos, 7–8.
Add MS 39683: [Thomas Cooke], *Observations upon Brewing, Fermentation, and Distillation, with sundry remarks and observations upon erection of corn distillhouses, situation, conveniences, repairs, expences, etc* [apparently unpublished, *c.* 1792?].

The British Museum, London, Department of Prints and Drawings

Trade cards referred to but not illustrated

Heal, 89.55: Anon., Trade card of Joseph Emerton, colour man (etching and engraving; verso: a manuscript bill made out to Charles Hayne Esqr., dated July 6 1744).
Heal, 68.99: Anon., Trade card for George Farr, grocer, at the Bee-hive and Three Sugar Loaves in Wood Street near Cheapside, London (etching and engraving, *c.* 1753).
Heal, 68.45: Murray, Trade card of R. Brunsden, tea dealer, grocer and oilman at the Three Golden Sugar Loaves in St James's Street (etching, 1750–60).
Heal, 36.38: Anon., Draft trade card of Robert Stone, nightman at the Golden Pole the upper end of Golden Lane near Old Street (1745).

Heal, 36.36: Anon., Draft trade card of Robert Stone, nightman and rubbish carter (1751).

Heal, 4.6: Anon., Draft trade card of Martha Prockter and Lydia Edwards, ass-keepers (verso: a bill dated 1775).

Banks, 44.11: Anon., Draft trade card of Benjamin Levy, coal merchant (engraving, *c.* 1780).

Banks, 44.25: Anon., Draft trade card J. Williams Junior, coal merchant, No.2 Beaumont's Buildings, St. George's East, and at Bush Wharf, Southwark (*c.* 1790).

Guildhall Library, London

Closed Access Broadside 30.74: S. T. Jannsen, *A Table Shewing the Number of Sheep and Black Cattle Brought to Smithfield Market for the Last Forty Years* (undated).

Sir John Soane's Museum, London

Adam Albums, Vol. 11, No. 33.

Lambeth Archives, London

P/S/13/19: Assessments for taxes on houses and windows; inhabited houses, male servants; four-wheeled carriages; riding and carriage horses; horses used in husbandry and trade; mules, 1800–01.

London Metropolitan Archives

New River Company records

ACC/2558/NR1/1–2: Minute books of weekly meetings and general courts 'A', 1769–78 and 1778–86.

ACC/2558/MW/C/15/341/010: Letter from Richard Cheffins to the board of directors, 3 April 1806.

Whitbread and Company Limited brewery records

4453/B/12/001–038: Rest books of Whitbread's brewery, Chiswell Street, 1799–1835.

Select bibliography

Truman Hanbury Buxton and Company Limited brewery records

B/THB/B/001–025/A: Rest books of the Black Eagle brewery, Spitalfields, 1741–1836.

Courage, Barclay & Simonds brewery records

ACC/2305/01/0834: A plan of an estate belonging to Henry Thrale esq. situate in Park Street, Maid Lane and Castle Lane, surveyed anno 1774 by George Gwilt, copy, photographic negative, 1983.
ACC/2305/01/0159/001: Rest book of Thrale's Anchor brewery, Southwark, 1780.
ACC/2305/01/0176/004: List of workers, 1791–97, a note in rest book of Barclay Perkins brewery, 1797.
ACC/2305/1/1300: Stable book of Barclay Perkins brewery, September 1827–April 1839.

Smithfield Market

CLA/016/FN/01/007: Rough weekly account: Tolls collected at Smithfield Market and the City Gates, 2 September 1727–28 September 1728.
CLA/016/FN/01/004: Dues collected at Smithfield Market, 28 March 1777–31 December 1817.
CLA/016/AD/01/003: Substance of the Bill Now Before Parliament, For Enlarging and Improving the Market Place with Observations Thereon: Also the Objections to the Proposed Measure, and Answer Thereto (1813).
CLA/016/AD/02/006: Second report of the Select Committee of the House of Commons appointed to inquire into the state of Smithfield Market, and the slaughtering of cattle in the metropolis (1828).
CLA/016/AD/02/011: Substance of the Cutting Butcher's Petition and Allegations Offered for an Alteration of Smithfield Market from Friday to Thursday (1796).
CLA/015/AD/02/032: Warrants for payments to constables and others for the apprehension and prosecution of persons, not being employed to drive cattle, for the 'hunting away' of bullocks, October–December 1789.

Sun Insurance Company

MS 11936–7: Policy registers, 1,262 volumes, 1710–1863.

Select bibliography

Court of Aldermen

COL/SP/05/084: Order of Court of Aldermen that persons who bring in stray cattle shall give their true name and address the Keeper of the Green-Yard and shall not be rewarded until 48 hours later; to prevent vagrant persons from driving cattle from fields to the Green Yard and giving fictitious names, 14 March 1731/2.

Middlesex Sessions Records

MJ/O/C/001: General orders of court, 6 December 1720.
MJ/O/C/006: General orders of court, 6 September 1753.
MR/L/SB/001: Registers of copies of slaughter house licences certified by ministers and churchwardens, 1786–1822.
MJ/SP/1755/09: Sessions Papers, 1755, Justices' working documents, September 1755.

Mansion House Justice Room

CLA/004/02/001–073: Minute books of the Mansion House Justice Room, City of London, 1784–1821.

Guildhall Justice Room

CLA/005/01/001–055: Minute books of the Guildhall Justice Room, City of London, 1752–1796.

London Consistory Court

DL/C/266 f. 142: Deposition of William Black, 9 June 1729.

Commissioners of Sewers and Pavements of the City of London

CLA/006/AD/04/004: Proceedings, 1766–97.

Holborn and Finsbury Commission of Sewers

HFCS/018: Minutes, 1763–98.

HFCS/011: Extracts from minutes, copies of warrants, etc., 1683–88.

Parish of Saint Saviour, Southwark

P92/SAV/0444: A survey and valuation of all the lands, buildings, houses, tenements and hereditaments within the parish of Saint Saviour Southwark by John Middleton and Thomas Swithin, surveyors, 1807–08.

Parish of Saint Mary Magdalene, Woolwich

P97/MRY/050: Vestry minutes, 1768–1819.

Diaries

ACC/1017/0944: Diary of John Eliot (III), 1757.

Parliamentary Archives, London

Islington to Paddington Road Bill

HL/PO/JO/10/3/250/4: Petition of Saint Andrews Holborn, Saint Georges Bloomsbury and Saint Giles' in the Fields, 6 April 1756.
HL/PO/JO/10/3/250/5: Petition of Saint George Hanover Square, St James Westminster, Saint Ann Soho, Paddington and Saint Marylebone, 6 April 1756.
HL/PO/JO/10/3/250/14: Petition of the Graziers, Salesmen, Butchers, Drovers and Dealers in Cattle who Attend Smithfield Market, 9 April 1756.

Royal Archives, Windsor

King's Mews

Mews/Proc/Mixed: Precedence book, 1760–1805.

Southwark Local History Library, London

2536: Vestry Minutes, Parish of St Giles, Camberwell, 29 October–5 November 1789.

Select bibliography

Sheffield Archives

Wentworth Woodhouse Muniments

WWM/A/1300: Weekly abstract, tradesmen's bills on house and stables, London, 1775–82.

WWM/A/1296: Isaac Charlton's London household disbursements, 19 July 1781.

WWM/A/1228: Inventory and appraisement, 1782–84.

WWM/A/1391/b: Discharges of Cash by Samuel Dutoit, 1766–67.

WWM/A/1300: Weekly abstract, tradesmens bills on house and stables, London, 1775–82.

Surrey History Centre, Woking

Surrey Quarter Sessions

QS2/6/1784/Mic/57: Sessions Bundles, Eleanor Harris, The Information of Joseph Whiting Holmes of the parish of Saint George Southwark… Cow Keeper, 19 July 1784.

The National Archives, Kew

Wills

PROB: Wills Proved at Prerogative Court of Canterbury.

Court of Chancery

C 5/240/16: Six Clerks Office, pleadings, Thomas Foxley and another (plaintiffs) v John Read and others (defendants), 1707.

Court of King's Bench

KB 10: Crown Side: London and Middlesex Indictment Files, 1675–1845.

Navy Board

ADM 110/9: Navy Board, Victualling Office, Out-letters, Letter Book, 1722–27.

ADM 112/162: Navy Board, Office of Surveyor of Victuals and Victualling Office, Contract Ledger, 1776.

Diary of Thomas Bridge (1760–1811)

J90/12–14.

Westminster City Archives, London

St Clement Danes parish records

B1074: St Clement Danes vestry minutes, 1796–1800.

St Martin-in-the-Fields parish records

F2505: Schedules of the sums assessed on the several wards in respect of duties upon shops, servants, horses, carriages and wagons, with the names of defaulters, 1785–98.

St Margaret Westminster parish records

E3159B: Duties on servants, carriages, horses and dog tax, commissioners' minute book, 1798–1804.

West Sussex County Records Office, Chichester

Petworth House Archives

PHA 6635: London tradesmen's bills from 25 June 1791 to 25 June 1792.
PHA 6640: Abstract of London tradesmen's bills paid for 3rd Earl of Egremont, 1794–95.
PHA 7555: Abstract, observations, and receipted bills from London tradesmen to 3rd Earl of Egremont, 1797–98.

Digitised sources (accessed 14 January 2019)

T. Hitchcock, R. Shoemaker, C. Emsley, S. Howard and J. McLaughlin., *The Old Bailey Proceedings Online, 1674–1913*, www.oldbaileyonline.org, version 8.0, March 2018.
T. Hitchcock, R. Shoemaker, S. Howard and J. McLaughlin., *London Lives, 1690–1800*, www.londonlives.org, version 2.0, March 2018.
17th and 18th Century Burney Collection Newspapers. Gale Cengage Learning.

Select bibliography

Published primary material

Published letters and diaries

Angelo, Henry, *Reminiscences of Henry Angelo with Memoirs of His Late Father and Friends*, 2 vols (1830).

Cozens-Hardy, Basil (ed.), *The Diary of Sylas Neville, 1767–1788* (London: Oxford University Press, 1950).

Grosley, Pierre Jean, *A Tour to London: Or, New Observations on England, and its Inhabitants. By M. Grosley, F.R.S. Member of the Royal Academies of Inscriptions and Belles Lettres*, trans. Thomas Nugent, 3 vols (Dublin, 1772).

Herbert, S. (ed.), *Pembroke Papers, 1780–1794: Letters and Diaries of Henry, Tenth Earl of Pembroke and his Circle* (London: Jonathan Cape, 1950).

Home, J. A. (ed.), *The Letters and Journals of Lady Mary Coke*,4 vols (Edinburgh, 1889–96).

Kemble, Frances Anne, *Records of a Girlhood*, 3 vols (New York, 1879).

Lewis, W. S. (ed.), *The Yale Edition of Horace Walpole's Correspondence*, 48 vols (New Haven: Yale University Press, 1937–83).

Lucas, Joseph (trans.), *Kalm's Account of his Visit to England on his way to America in 1748* (London, 1892).

Maxwell, H. (ed.), *The Creevey Papers: A Selection From the Correspondence and Diaries of the Late Thomas Creevey, MP*, 2 vols (Cambridge: Cambridge University Press, 2012).

Pictet, Marc-Auguste, *Voyage de Trois Mois, en Engleterre, en Écosse et en Irelande* (Geneva, 1802).

Savile, Alan (ed.), *Secret Comment, The Diaries of Gertrude Savile, 1721–1759* (Nottingham: Thoroton Society, 1997).

Simond, Louis, *Journal of a Tour and Residence in Great Britain, During the Years 1810 and 1811*, 2 vols (Edinburgh, 1817).

Sturge, Clement Young (ed.), *Leaves from the Past: The Diary of John Allen, Sometime Brewer of Wapping 1757–1808* (Bristol: Arrowsmith, 1905).

Thale, Mary (ed.), *The Autobiography of Francis Place, 1771–1854* (Cambridge: Cambridge University Press, 1972).

Walpole, Horace, *The Letters of Horace Walpole, Earl of Orford: Including Numerous Letters now First Published from the Original Manuscripts*, ed. [anon.], 6 vols (1840).

Wood, Florence and Wood, Kenneth (eds), *A Lancashire Gentleman: The Letters and Journals of Richard Hodgkinson, 1763–1847* (Stroud: Allan Sutton Press, 1992).

Select bibliography

Microfilmed Collections

The Boulton & Watt Archive and the Matthew Boulton Papers from the Birmingham Central Library (Adam Matthew Publications).

Newspapers and Periodicals

Common Sense or The Englishman's Journal
Country Journal or The Craftsman
Daily Advertiser
Daily Journal
Gazetteer & New Daily Advertiser
The Gentleman's Magazine
London Chronicle
London Daily Advertiser
London Daily Post
Lloyd's Evening Post
The Monthly Magazine, and British Register
Morning Chronicle & London Advertiser
Morning Post & Daily Advertiser
Morning Post & Fashionable World
Morning Post & Gazetteer
Oracle & Daily Advertiser
Oracle & Public Advertiser
Philosophical Transactions of the Royal Society of London
The Repertory of Arts and Manufactures
Spectator
Sporting Magazine or Monthly Calendar
St James's Chronicle
The Universal Magazine
The Westminster Magazine
Whitehall Evening Post
World

Maps

Horwood, Richard, *Plan of the Cities of London and Westminster, with the Borough of Southwark and parts adjoining showing every house* (3rd edn, 1813), republished as *The A to Z of Regency London* (London: London Topographical Society, 1985).

Potter, Peter, *Plan of the Parish of St Marylebone in the County of Middlesex* (2nd edn, 1821).

Rocque, John, *A Plan of the Cities of London and Westminster, and Borough of Southwark, London* (1747), republished as *The A to Z of Georgian London* (London: London Topographical Society, 1982).

Engravings (referenced but not illustrated)

Anon., *Interior View of the Auction Rooms at Aldridge's Horse Repository* (Lithograph, 1824).

Bowles III, Thomas, *A New and Exact Prospect of the North Side of the City of London taken from the Upper Pond near Islington* (etching and engraving, 1730).

Cruikshank, Isaac, *Effects of the Dog Tax* (hand-coloured etching, 1796).

Dighton, Robert, *Two Impures of the Ton driving to the Gigg Shop, Hammersmith* (Mezzotint, 1781).

Egan, James, after John Christian Zeitter, *Pirate and Outlaw* (etching and engraving, *c.* 1818).

Gilray, James, *A Pig in a Poke. Whist, Whist* (hand-coloured etching, 1788)

— *Pigs Meat; – or – The Swine Flogg'd Out of the Farm Yard* (hand-coloured etching, 1798).

Gravelot, H., *A Perspective View of the Engine, Now Made Use of for Driving the Piles of the New Bridge at Westminster* (etching and engraving, 1748).

Lane, Theodore, 'How to Pick up a "RUM ONE to Look at" and a "GOOD ONE to Go" in Smithfield' (hand-coloured etching and aquatint, 1825).

Rowlandson, Thomas, *The Light Horse Volunteers of London & Westminster, Commanded by Coll Herries, Reviewed by His Majesty on Wimbledon Common, 5th July 1798* (etching and aquatint, 1798).

— *Easter Monday, or the Cockney Hunt* (hand-coloured etching, 1811).

White, Charles, after anon., *View of the River Thames from the End of Chelsea* (engraving, 1794).

Wolstenholme the elder, Dean, *Red Lion Brewhouse, East Smithfield* (mezzotint, 1805).

— *A Correct View of the Golden Lane Genuine Brewery* (mezzotint, 1807).

— *A Correct View of the Hour Glass Brewery Belonging to Messrs Calvert and Co.* (coloured aquatint, 1821).

Select bibliography

Parliamentary proceedings

Journal of the House of Commons, vol. 23: 1737–1741 (1803).
Journal of the House of Commons, vol. 24: 1741–1745 (1803).
Journal of the House of Commons, vol. 61: 1806 (1806).
The Parliamentary Debates from the Year 1803 to the Present Times, vol. 26 (1812).
The Parliamentary History of England, from the Earliest Period to the Year 1803, vol. 31 (1818) and vol. 32 (1818).
The Parliamentary Register; or History of the Proceedings and Debates of the House of Commons, vol. 44 (1796).

Other (place of publication London unless otherwise stated)

Adams, John, *Analysis of Horsemanship: Teaching the Whole Art of Riding in the Manege, Military, Hunting, Racing, or Travelling System* (Edinburgh, 1799).
Adams, William Bridges, *English Pleasure Carriages: Their Origin, History, Varieties, Materials, Construction, Defects, Improvements, and Capabilities* (1837).
Allen, Thomas, *The History and Antiquities of London, Westminster and Southwark, and Parts Adjacent*, 4 vols (1827–29).
Allen, Zachariah, *Sketches of the State of the Useful Arts, and of Society, Scenery, &c. &c. in Great Britain, France and Holland. Or, the Practical Tourist*, 2 vols (Boston, 1833).
Anon., *The Red Cow's Speech, to a Milk-Woman, in St. J----S's P----K* (early 1700s).
— *An Impartial Enquiry into the Present State of the British Distillery; Plainly Demonstrating the Evil Consequences of Imposing any Additional Duties on British Spirits* (1736).
— *A Dissertation on Mr. Hogarth's Six Prints Lately Published, viz. Gin-Lane, Beer-Street, and the Four Stages of Cruelty* (1751).
— *Eclipse Races (Addressed to the Ladies:) Being an Impartial Account of the Celestial Coursers and their Riders… By Philo-Pegasus, a Lover of Truth* (1764).
— *The Complete Grazier: or, Gentleman and Farmer's Directory* (2nd edn, 1767).
— *The Corn Distillery, Stated to the Consideration of the Landed Interest of England* (1783).
— *The Lamentation of a Dog, on the Tax, and its Consequences. Addressed to the Right Hon. William Pitt. With Notes by Scriblerus Secundus* (1796).
— *A Treatise on Milk, as an Article of the First Necessity to the Health and Comfort of the Community: A Review of the Difference Methods of Production; and*

Suggestions Respecting the Best Means of Improving its Quality, Reducing its Price, and Increasing its Consumption (1825).

— *How to Live in London; or, the Metropolitan Microscope, and Stranger's Guide* (1828).

Baird, Thomas, *General View of the Agriculture of the County of Middlesex: With Observations on the Means of its Improvement* (1793).

Barfoot, Peter and Wilkes, John, *The Universal British Directory of Trade, Commerce, and Manufacture*, 5 vols (1790–98).

Barry, Edward, *On the Necessity of Adopting Some Measures to Reduce the Present Number of Dogs; with a Short Account of Hydrophobia, and the Most Approved Remedies Against it, a Letter, to Frances Annesley, Esq., MP for the Borough of Reading, and one of the Trustees of the British Museum* (Reading, 1796).

Bath and West of England Society, *Letters and Papers on Agriculture, Planting &c., Selected from the Correspondence of the Bath and West of England Society, for the Encouragement of Agriculture, Arts, Manufactures and Commerce*, vol. 7 (Bath, 1795).

Bell, James, *A System of Geography, Popular and Scientific, or a Physical, Political, and Statistical Account of the World and its Various Divisions*, 6 vols (Glasgow, 1832).

Bell, Thomas, *A History of British Quadrupeds, Including the Cetacea* (1837).

Berenger, Richard, *The History and Art of Horsemanship*, 2 vols (1771).

Blackstone, William, *Commentaries on the Laws of England*, ed. Wilfrid Prest, 4 vols (Oxford, 1765–69; Oxford: Oxford University Press, 2016).

Boswell, James, *The Journal of a Tour to the Hebrides, with Samuel Johnson, LL.D.* (Dublin, 1785).

— *The Life of Samuel Johnson, LL.D.* (1791; Ware: Wordsworth Editions, 2008).

Bracken, Henry, *Farriery Improved; or, a Complete Treatise on the Art of Farriery* (12th edn, 1792).

Bradley, Richard, *The Gentleman and Farmer's Guide for the Increase and Improvement of Cattle* (2nd edn, 1732).

— *A General Description of All Trades, Digested in Alphabetical Order* (1747).

Brewster, David (ed.), *The Edinburgh Encyclopaedia*, 18 vols (Philadelphia, 1832).

Brown, Joseph, *The Circus; or, British Olympicks. A Satyr on the Ring in Hyde-Park* (1709).

Bunbury, Henry, *An Academy for Grown Horsemen, Containing the Completest Instructions for Walking, Trotting, Cantering, Galloping, Stumbling, and Tumbling* (1787).

Campbell, Robert, *The London Tradesman. Being a Compendious View of All the Trades, Professions, Arts, Both Liberal and Mechanic, Now Practiced in the Cities of London and Westminster* (1747).

Caulfield, James, *Blackguardiana: Or, a Dictionary of Rogues, Bawds, Pimps, Whores, Pickpockets, Shoplifters* (1793).

Cobbett, William, *Rural Rides*, ed. Ian Dyck (1830; Harmondsworth: Penguin, 2001).

Colquhoun, Patrick, *A Treatise on the Police of the Metropolis: Containing a Detail of the Various Crimes and Misdemeanours by which Public and Private Property and Security are, at Present, Injured and Endangered: and Suggesting Remedies for Their Prevention* (4th edn, 1797).

Cook, John, *Observations on Fox Hunting, and the Management of Hounds in the Kennel and the Field* (1826).

[Cox, Thomas], *Magna Britannia Antiqua & Nova: Or, a New, Exact, and Comprehensive Survey of the Ancient and Present State of Great-Britain*, 6 vols (1738).

Culley, George, *Observations on Live Stock, Containing Hints for Choosing and Improving the Best Breeds of the Most Useful Kinds of Domestic Animals* (1786).

Curtis, Thomas (ed.), *The London Encyclopaedia or Universal Dictionary of Science, Art, Literature and Practical Mechanics*, 22 vols (1829).

Curwen, John Christian, *Hints on Agricultural Subjects and on the Best Means of Improving the Condition of the Labouring Classes* (2nd edn, 1809).

Defoe, Daniel, *A Tour through the Whole Island of Great Britain*, ed. Pat Rogers (1724–26; Harmondsworth: Penguin, 1986).

Dibdin, Charles, *The High Mettled Racer* (1831).

Diderot, Denis and D'Alembert, Jean Le Rond (eds), *Encyclopédie ou Dictionnaire Raisonné des Arts et des Métiers*, 28 vols (Paris, 1751–72).

Dodd, George, *The Food of London: A Sketch of the Chief Varieties, Sources of Supply, Probable Quantities, Modes of Arrival, Processes of Manufacture, Suspected Adulteration, and Machinery of Distribution, of the Food for a Community of Two Millions and a Half* (1856).

Egan, Pierce, *Life in London: or, the Day and Night Scenes of Jerry Hawthorn, Esq. and his Elegant Friend Corinthian Tom, Accompanied by Bob Logic, the Oxonian, in their Rambles and Sprees through the Metropolis* (1821).

Ellis, William, *The Second Part of the Timber-Tree Improved* (1742).

— *The London and Country Brewer* (5th edn, 1744).

— *A Compleat System of Experienced Improvements, made on Sheep, Grass-Lambs, and House- Lambs: Or, The Country Gentleman's, the Grasier's, the Sheep-Dealer's, and The Shepherd's Sure Guide* (1749).

Elmes, James, *Metropolitan Improvements: Or, London in the Nineteenth Century, Displayed in a Series of Engravings of the New Buildings, Improvements, &c. by the Most Eminent Artists, from Original Drawings, taken from the Objects themselves expressly for this Work* (1827–30).

Farey, John, *A Treatise on the Steam Engine, Historical, Practical, and Descriptive* (1827).

Feltham, John, *The Picture of London* (1807, 1813, 1818, 1825).

Foot, Peter, *General View of the Agriculture of the County of Middlesex, with Observations on the Means of their Improvement* (1794).

Goldsmith, Oliver, *An History of the Earth, and Animated Nature*, 8 vols (1774).

Gordon, W. J., *The Horse World of London* (1893).

Gregory, George, *A New and Complete Dictionary of Arts & Sciences*, 3 vols (New York, 1819).

Grose, Francis, *A Classical Dictionary of the Vulgar Tongue* (1785).

Gwynn, John, *London and Westminster Improved, Illustrated by Plans: To which is Prefixed, a Discourse on Publick Magnificence* (1766).

Hanway, Jonas, *A Letter to Mr. John Spranger on his Excellent Proposal for Paving, Cleansing, and Lighting the Streets of Westminster, and the Parishes in Middlesex* (1754).

Houghton, John, *Husbandry and Trade Improv'd: Being a Collection of Many Valuable Materials Relating to Corn, Cattle, Coals, Hops, Wool &c.*, revised by Richard Bradley, 3 vols (1727).

Hughson, David, *London; Being an Accurate History and Description of the British Metropolis and its Neighbourhood: To Thirty Miles Extent, from an Actual Perambulation*, 6 volumes (1809).

James, William and Malcolm, Jacob, *General View of the Agriculture of the County of Surrey* (1794).

Jenner, Charles, *Town Eclogues* (1772).

King, William, *Reasons and Proposals for Laying a Tax upon Dogs. Humbly Addressed to the Honourable House of Commons* (Reading, 1740).

[Legg, Thomas], *Low-Life; or One Half of the World Knows Not How the Other Half Live… In the Twenty-four Hours, Between Saturday-Night and Monday-Morning. In a True Description of a Sunday, as it is Usually Spent Within the Bills of Mortality* (3rd edn, 1764).

Lawrence, John, *A Philosophical and Practical Treatise on Horses, and on the Moral Duties of Man Towards the Brute Creation*, 2 vols (1796–98).

— *The Horse in All His Varieties and Uses; His Breeding, Rearing, and Management, Whether in Labour or Rest; with Rules, Occasionally Interspersed, for His Preservation from Disease* (1829).

[Lawrence, John], *The Sportsman's Repository; Comprising a Series of Highly-Finished Engravings, Representing the Horse and the Dog, in All Their Varieties;*

Executed in the Line Manner, by John Scott, from Original Paintings by Marshall, Reinagle, Gilpin, Stubbs, and Cooper... (1820).

Lysons, Daniel, *The Environs of London: Being an Historical Account of the Towns, Villages, and Hamlets, within Twelve Miles of that Capital*, 4 vols (1795–96).

Maitland, William, *The History and Survey of London from its Foundation to the Present Time* (1 vol., 1st edn, 1739; 2 vols, 2nd edn, 1756; 2 vols, 3rd edn, 1760).

Marshall, William, *Minutes of Agriculture; with Experiments and Observations Concerning Agriculture and the Weather* (1783).

Martin, Benjamin, *The General Magazines of Arts and Sciences, Philosophical, Philological, Mathematical, and Mechanical*, 14 vols (1755–65).

[Massie, Joseph], *An Essay on the Many Advantages Accruing to the Community from the Superior Neatness, Conveniences, Decorations and Embellishments of Great and Capital Cities* (1754).

Matthews, William, *Hydraulica, an Historical and Descriptive Account of the Water Works of London: and the Contrivances for Supplying Other Great Cities, in Different Ages and Countries* (1835).

Mayhew, Henry, *London Labour and the London Poor. Vol. 1. The London Street-Folk* (1851).

— *London Labour and the London Poor*, 4 vols (1861–62).

Middleton, John, *View of the Agriculture of Middlesex; with Observations on the Means of its Improvement, and Several Essays on Agriculture in General. Drawn up for the Consideration of the Board of Agriculture* (1798).

Nelson, John, *The History, Topography, and Antiquities of the Parish of St. Mary Islington, in the County of Middlesex* (1811).

Nightingale, Joseph, *London and Middlesex: or, an Historical, Commercial, & Descriptive Survey of the Metropolis of Great-Britain: Including Sketches of its Environs, and a Topographical Account of the Most Remarkable Places in the Above County... Vol. 3* (1815).

Pennant, Thomas, *Of London* (1790).

Pigot & Co., *London and Provincial New Commercial Directory for 1826–27* (1826).

— *Metropolitan New Alphabetical Directory for 1827* (1827).

Pyne, William Henry, *The Costume of Great Britain* (1804).

Smollett, Tobias, *The Expedition of Humphry Clinker*, ed., Lewis M. Knapp, revised by Paul-Gabriel Boucé (1771; Oxford: Oxford University Press, 1998).

[Stephen, George], *The Adventures of a Gentleman in Search of a Horse, by Caveat Emptor* (1835).

Stowe, John [updated by John Strype], *A Survey of the Cities of London and Westminster: Containing the Original, Antiquity, Increase, Modern Estate and*

Government of those Cities Written at first in the Year MDXCVIII by John Stow, Citizen and Native of London, 2 vols (1720).

Surtees, Robert Smith, *Jorrocks's Jaunts and Jollities: The Hunting, Shooting, Racing, Driving, Sailing, Eating, Eccentric and Extravagant Exploits of that Renowned Sporting Citizen Mr. John Jorrocks of St. Botolph Lane and Great Coram Street* (2nd edn, 1843).

Swift, Jonathan, *Directions to Servants in General* (1745; London: Hesperus Press, 2003).

Taplin, William, *The Gentleman's Stable Directory; or, Modern System of Farriery*, 2 vols (1788).

— *The Sporting Dictionary, and Rural Repository of General Information Upon Every Subject Appertaining to the Sports of the Field*, 2 vols (1803).

Trusler, John, *The London Advisor and Guide: Containing Every Instruction and Information Useful and Necessary to Persons Living in London, and Coming to Reside There* (1786).

[Tucker, Josiah], *An Impartial Inquiry into the Benefits and Damages Arising to the Nation from the Present Very Great Use of Low-priced Spiritous Liquors: With Proper Estimates Thereupon, and Some Considerations Humbly Offered for Preventing the Introduction of Foreign Spirits Not Paying the Duties. By J.T. of Bristol* (1751).

Wilson, Thomas, *Distilled Spiritous Liquors the Bane of the Nation: being some considerations humbly offer'd to the legislature* (1736).

Wordsworth, William, *The Prelude or Growth of a Poet's Mind (Text of 1805)*, ed., Ernest de Selincourt, corrected by Stephen Gill (Oxford: Oxford University Press, 1970).

Young, Arthur (ed.), *Annals of Agriculture and Other Useful Arts*, 46 vols (1784–1815).

Youatt, William, *The Horse; With a Treatise on Draught and a Copious Index* (1831).

— *Cattle; Their Breeds, Management, and Diseases* (1834).

Index

Page references in *italics* indicate illustrations. Page references for notes give the page reference and the note number in the format 239n31.

Index

Index

Index

Index

Index

Haverstock Hill
Spring Pla.
Chalk-house
Copenhagen House
Chalk Farm
Back Road to Highgate
Pancras Work-house
Mother Red Caps
Chap.
Regents Canal
CAMDEN TOWN
Veterinary College
Pancras new Work house
PANCRAS
White Conduit Ho.
Islington Work-house
I S L I N G
Reservoir
THE REGENTS PARK
Prince of Wales Circus
PENTON VILLE
SOMERS TOWN
Spa Fields
Bagnigge Wells
Grosvenor Gate
Reservoir
HYDE PARK
Green Park
ST. JAMES'S PARK
Constitution Hill
Hyde Park Cor.
Queens Garden
Cadogan Square
Hans Squ.
Wire Manuf.
Pimlico
Bridewell
Tothill Fields
New Penitentiary
Water Works
Willow Walk
Neat Houses Gardens
Hospital
Chelsea Water Works
CHELSEA REACH
Nine Elms
Vauxhall Gardens
L A M B E T H
NEWINGTON
Manor Ponds
Kennington Common
Furlongs
LON
Published as the Act directs